# Design by Fire

Across the world, the risks of wildfires are increasing and expanding. Due to human actions, we dwell in the age of fire – the Pyrocene – and the many challenges and climate adaptation questions it provokes. Exploring our past and current relationships with fire, this book speculates on the pyro futures yet to be designed and cared for.

Drawing upon fieldwork, mapping, drone imagery, and interviews, this publication curates 27 global design case studies within the vulnerable and dynamic wildland-urban interface and its adjacent wildlands. The book catalogs these examples into three approaches: those that resist the creative and transformative power of fire and forces of landscape change, those that embrace and utilize those forces, and those that intentionally try to retreat and minimize human intervention in fire-prone landscapes. Rather than serving as a book of neatly packaged solutions, it is a book of techniques to be considered, tested, and evaluated in a time of fire.

**Emily Schlickman** is an Assistant Professor of Landscape Architecture and Environmental Design at the University of California, Davis whose research explores design techniques for accelerated climate change. Schlickman received a BA from Washington University in St. Louis and an MLA from Harvard University Graduate School of Design.

**Brett Milligan** is a landscape architect and Professor of Landscape Architecture and Environmental Design at the University of California, Davis. There he is the director of the Metamorphic Landscapes Lab, dedicated to prototyping landscape-based adaptations to conditions of accelerated climatic and environmental change, through extensive fieldwork and transdisciplinary design research. Much of his work is based in California, undoing and reworking colonial legacies of land reclamation, water infrastructure, flood control, and fire suppression. He is a co-founder of the Dredge Research Collaborative, a nonprofit organization dedicated to exploring the human alteration and design of sedimentary landscapes, and co-author of the book *Silt Sand Slurry: Dredging, Sediment, and the Worlds We Are Making*.

"It's no longer enough to live with fire. We have to live with a fire age. That requires new thinking, novel classifications, fresh metaphors and models, a vision of what can happen where fire, town, and country converge, so it's great to see what landscape architects have to say. *Design by Fire* is a welcome contribution to an urgent problem."

**Stephen Pyne**, ASU, author of *The Pyrocene*

"*Design by Fire* is the essential guidebook and atlas for the pyro-future that is already here. Whether homeowner, concerned citizen, designer, or policymaker, you will find in these extraordinarily researched and illustrated pages a foundation for understanding – and living in – the world to come."

**Alexander Robinson**, USC School of Architecture, author of *The Spoils of Dust: Reinventing the Lake that Made Los Angeles*

"*Design by Fire* is a necessary book for all landscape architects and planners. The insightful interviews, succinct strategies, and emphasis on co-creative approaches structure the book, while the authors challenge us to grapple with current practices. They help us imagine a future in which reparations can bring traditional ecological knowledge to the forefront and imbue us with a culture of stewardship."

**Miho Mazereeuw**, MIT, Director of The Urban Risk Lab

# Design by Fire

## Resistance, Co-Creation, and Retreat in the Pyrocene

Emily Schlickman and Brett Milligan

NEW YORK AND LONDON

Cover image: View of a landscape and highway around Lake Berryessa after the LNU Lightning Complex Fires of 2020. Photograph by Brett Milligan.
Cover design: Emily Schlickman.

First published 2023
by Routledge
605 Third Avenue, New York, NY 10158

and by Routledge
4 Park Square, Milton Park, Abingdon, Oxon, OX14 4RN

*Routledge is an imprint of the Taylor & Francis Group, an informa business*

© 2023 Emily Schlickman and Brett Milligan

*Library of Congress Cataloging-in-Publication Data*
Names: Schlickman, Emily, author. | Milligan, Brett, author.
Title: Design by fire: resistance, co-creation and retreat in the Pyrocene /
Emily Schlickman and Brett Milligan.
Description: New York, NY: Routledge, 2023. |
Includes bibliographical references and index. |
Identifiers: LCCN 2022061763 (print) | LCCN 2022061764 (ebook) |
ISBN 9781032001494 (hardback) | ISBN 9780367767617 (paperback) |
ISBN 9781003172956 (ebook)
Subjects: LCSH: Firescaping—Case studies. | Landscape design—Case studies.
Classification: LCC SB475.9.F57 S35 2023 (print) | LCC SB475.9.F57
(ebook) | DDC 635.9/5—dc23/eng/20230412
LC record available at https://lccn.loc.gov/2022061763
LC ebook record available at https://lccn.loc.gov/2022061764

ISBN: 9781032001494 (hbk)
ISBN: 9780367767617 (pbk)
ISBN: 9781003172956 (ebk)

DOI: 10.4324/9781003172956

Typeset in Univers
by codeMantra

Figure 0.1
Aerial view of a conifer grove after a wildfire event with limited tree survival.
Photograph by Derek Young.

Figure 0.2
View of a conifer grove after a wildfire event with many trees not surviving.
Photograph by Derek Young.

Figure 0.3
Aerial view of a landscape around Lake Berryessa after the LNU Lightning Complex Fires
of 2020.

Figure 0.4
View of a valley near Lake Berryessa after the LNU Lightning Complex Fires of 2020.

Figure 0.5
View of a conifer grove after a wildfire event showing the extent of damage.
Photograph by Derek Young.

Figure 0.6
View of a landscape and highway around Lake Berryessa after the LNU Lightning
Complex Fires of 2020.

Figure 0.7
Aerial view of a conifer grove after a wildfire event with limited tree survival.
Photograph by Derek Young.

*For all fire stewards around the world, and for those yet to be.*

# Contents

# Introduction

Stewarding Change

Figure 1.1

View of a landscape around Lake Berryessa after the LNU Lightning Complex Fires of 2020.

# Chapter 1

# Stewarding Change

*The worldview of a society is often written more truthfully on the land than in its documents.*[1]

Robin Wall Kimmerer and Frank Kanawha Lake

*The touchstone for Wilderness turns out to be an artifact of generations of human care.*[2]

Rebecca Solnit

## A Valley Remade

In 1865, Frederick Law Olmsted stood below the towering granite cliffs of El Capitan and read aloud his preliminary report to fellow appointed commissioners charged with the planning and creation of Yosemite Park. Above all, Olmsted called for the strict preservation of a landscape he deemed to be a pristine wilderness:

> *The first point to be kept in mind then is the preservation and maintenance as exactly as is possible of the natural scenery; the restriction, that is to say, within the narrowest limits consistent with the necessary accommodations of visitors, of all artificial constructions and the prevention of all constructions markedly inharmonious with the scenery or which would unnecessarily obscure, distort or detract from the dignity of the scenery.*[3]

In seeking to *preserve* this scenery, his report denounced the presence and use of fire within Yosemite. Olmsted remarked that "Indians and others have set fire to the forests and herbage and numbers of trees have been killed by these fires," and further, "not only have trees been cut, hacked, barked and fired in prominent positions, but rocks in the midst of the most picturesque natural scenery have been broken, painted and discolored, by fires built against them."[4] For Olmsted, *natural scenery* – that which was not human, or human modified – was the primary reason for the congressional appropriation of these lands for its citizens; in his words, the very "peculiarity of this ground"[5] prompting its preservation. As a measure of emphasis, Olmsted evokes the word *scenery* 33 times in his preliminary report, claiming that natural scenery is a provider of "pecuniary" gains, a source of public health and recovery from the tolls of urban living, and a means for cultural refinement.

DOI: 10.4324/9781003172956-2

Generally speaking, scenery refers to the features and visual qualities of a land-scape; of how a landscape appears as a setting or relatively fixed background. Like theater, it is the stage upon which experience is guided. And it is clear from the commissioner's report that this is how Olmsted meant it. As a landscape architect, his aesthetic concern was focused on artistic appreciation and how visual perception could be culturally elevated beyond a baseline of mere savagery:

> *The power of scenery to affect men is, in a large way, proportionate to the degree of their civilization and to the degree in which their taste has been culti-vated. Among a thousand savaged savages there will be a much smaller number who will show the least sign of being so affected than among a thousand persons taken from a civilized community.*[6]

The opening passages of Olmsted's preliminary report are composed of lengthy and poetic descriptions of Yosemite's sublime "chasm." He describes the s*cenery* at length, and per the times, sees much of it through a Eurocentric lens:

> *There is nothing strange or exotic in the character of the vegetation; most of the trees and plants, especially of the meadow and waterside, are closely allied to and are not readily distinguished from those most common in the landscapes of the Eastern States or the midland counties of England. The stream is such a one as Shakespeare delighted in, and brings pleasing reminiscences to the traveller of the Avon or the Upper Thames…in following up the ravines, cabinet pictures open at every turn, which, while composed of materials mainly new to the artist, constantly recall the most valued sketches of Calame in the Alps and Apennines.*[7]

Olmsted dedicates the closing arguments of his report to advance Yosemite's "present condition as a museum of natural science." Yet, no ecological specifics or descrip-tion appears in the text, nor what "natural" science actually means. In terms of physical park design, Olmsted's primary design move was infrastructural: to build a new road to access the park more easily, and for the Mariposa Grove in particular, to have this road be:

> *carried completely around it, so as to offer a barrier of bare ground to the approach of fires, which nearly every year sweep upon it from the adjoining country, and which during the last year alone have caused injuries, exemption from which it will be thought before many years would have been cheaply obtained at ten times the cost of the road.*[8]

Here again, we see that from the very inception of the park, there was a clear intent to banish fire, which was understood in destructive terms set in opposition to preservation goals. Time would prove this to be an erroneous understanding of California's montane landscapes that would have profound consequences.

Interestingly, Olmsted's report never made it beyond the valley and its recommen-dations were largely ignored. And having failed, Olmsted resigned from the commission the following year. Yet, his call for fire suppression was widely popular and adopted as a man-agement practice for Yosemite Park and the Western U.S. generally.

Olmstead did return to Yosemite 24 years later, and noted that the landscape had significantly changed. Cattle pastures had replaced meadows; tourist lodging had been carved out of majestic groves; and dense new understories, unchecked by fire, had obscured many of his cherished scenic views across the valley. Olmsted published his critical views of the park in a 1890 pamphlet, in which he dwells again upon the topic of "artificiality," claiming it has no rightful place in this wilderness: "nothing of an artificial character should be allowed a place on the property, no matter how valuable it might be under other circumstances and no matter how little cost it may be had."[9] And while many of the changes that had come to his beloved landscape were commercially exploitive and transformative, Olmsted, like most others, failed to recognize the loss of the most crucial phenomena in sustaining the scenery of the Yosemite that he loved: fire.[10]

As many know, the landscape that Olmsted stepped into in 1865 was far from untouched, pristine wilderness. Rather, it was a highly tended landscape on the grandest of scales. For generations, the Ahwahnechee, the indigenous people of the valley they called *Ahwahnee*, intentionally set fires and allowed lightning to burn the fire-adapted lands. These routine practices would reset successional processes and promote the growth of indigenous vegetation they valued, including oaks, bunch grass, sedge root, and milkweed – plants that were then used for medicinal purposes, food production, and as raw materials for everyday items and shelter.[11]

Olmsted was not alone among wilderness advocates and leading conservationists in erasing the co-creative role of Native Americans in Yosemite's landscapes and elsewhere. Kat Anderson, a prominent scholar of traditional ecological knowledge in California, has observed that John Muir, Olmsted's contemporary and the influential founder of the Sierra Club,

> …*was an early proponent of the view that the California landscape was a pristine wilderness before the arrival of Europeans. Staring in awe at the lengthy vistas of his beloved Yosemite Valley, or the extensive beds of golden and purple flowers in the Central Valley, Muir was eyeing what were really the fertile seed, bulb, and greens gathering grounds of the Miwok and Yokuts Indians, kept open and productive by centuries of carefully planned indigenous burning, harvesting, and seed scattering.*[12]

The fires of pre-colonial Yosemite, whether ignited by lighting or people, tended to burn relatively mild and cool, given in part to their designed frequency. In particular, intentionally lit fires were timed to be small and lower intensity, behaving like massive roving herbivores, primarily impacting the surface of the ground and leaving behind nutrients and charred organic matter to enrich soils.[13] Thus when wildfires did occur (such as by lightning), they might burn at a range of intensities, but overall, tended to burn in less severe ways, given the cultural burns that kept fuel levels reduced. The Ahwahnechee used these tactics to sculpt and steward the landscape of Yosemite until 1851, when Major John Savage of the Mariposa Battalion and his militia forcefully pushed them out of the valley, ironically using embers from their fire pits to ignite the camp.[14] After that, indigenous cultural burning in the region essentially stopped.

Olmsted stood in Yosemite pronouncing his vision for the park just 14 years later. In calling for fire eradication, Olmsted must have been unaware of fundamental contradictions ingrained in his proposal. He had adamant contempt for anything "artificial" within the park,

yet his fire suppression agenda and its implementation for over a century, became a far grander artifice in the valley than the roads, cabins, skating rinks, and other changes brought by the hordes of visitors to the park. Yosemite morphed into a very different landscape because of fire expulsion; a radical shift from the thousands of years of human-fire-landscape coevolution that preceded it.

Here, we cannot help but collide with the unreal dichotomies of Olmsted's *natural* and *artificial*. With respect to Yosemite, Olmsted's understanding of *natural* meant a lack of human intervention and *artifice*; of nature as a pristine, other-than-human background; in his words, "the prevention of all constructions markedly inharmonious with the scenery or which would unnecessarily obscure, distort or detract from the dignity of the scenery."[15] Yet, the human *artifice* of fire suppression is what remade Yosemite into an entirely different *scene* despite what Olmsted and many after him so Sysiphusly tried to preserve in static form. The fires that Olmsted banished, due to a lack of understanding of fire-dependent Western U.S. landscapes, was the very thing that could have held its "natural" scenery to some semblance of what it was across time. More foundational than its changed looks, fire suppression changed the vegetative structure, hydrology, and ecological functioning of Yosemite and so many other landscapes in California and elsewhere. We say that Olmsted's exclusive oppositions of *natural* and *artificial* are unreal because each actually means and makes the other, inseparably. Additionally, the untenable understanding of nature as non-human, helped promote the "pristine myth" of the West as a misleading and colonizing instrument in Olmsted's days.[16] To paraphrase William Cronon, there is nothing natural about the Western conception of wilderness.[17] And today, as we observe and feel the long-term effects of land management premised on that approach, that understanding of *nature* is rendered as culturally and empirically absurd.

Ironically, Olmsted's forgotten preliminary report resurfaced over a century after he wrote it. It was brought back into circulation by his biographer to be hailed as a work of wilderness conservation planning ahead of its time.[18] It is praised for resisting private development and the rampant exploitation of wild lands at the expense of public quality of life.[19] But that praise needs tempering, as Olmsted was also part of his cultural and colonial milieu and its missteps. His assumptions about landscape preservation and the assumed ability to maintain landscapes in a static form, is also exemplary of planning, engineering, and design paradigms of the 19th and early 20th centuries, the detrimental legacies of which challenge us today. Olmsted's loose comparisons to vegetation found in the vastly different climate and habitats of England evidences a surficial, visual abstraction of landscapes at the expense of deeper and place-specific understandings of California's Sierra Nevada Mountains. Olmsted did not seem to understand or *see* the entwined eco-cultural processes that had created what he was aestheticizing. He actually spent very little time in these landscapes, and in private correspondences, expressed his personal unease with California's wilderness generally.[20] It also seems fair to say that Olmsted was either unaware or accepting of his role as an instrument of the state, in advancing its colonial practices in the recently claimed Western U.S. The publics he advocates for are clearly those of settlers and not the people that were native to it. The inhabitants who tarnished the color and purity of Yosemite's boulders with the dark, ashen powders of their fires – per his description – were clearly not to be included in this park. There are many things that Olmsted got very wrong in his report, which opens up many questions regarding what might have happened had the report been

Figure 1.2 (*Left*) Extent and density of forested areas in Yosemite National Park in pre-colonial times, during which cultural burns were routinely applied to shape vegetative communities in the valley (top) and the same view after a century of Euro-American fire suppression (bottom).

researched and approached differently, as well as questions about what stories and lessons we tell about landscape architecture's disciplinary legacies. And clearly, from Olmsted we can see how much aesthetic sensibilities and the types of experiences, ideas, and cultural norms they are based on, matter profoundly.

## *Rise of the Feral*

In many ways, Yosemite has indeed become "a museum of natural science," but not in the manner Olmsted had intended it. Much recent scientific research and management practices in Yosemite have sought to understand exactly how fire suppression transformed these lands, rather than preserving them. The museum of what was can be found in the park's tree rings, in pollen preserved in its lake beds, in soil profiles, and in remnant carbon-based materials. It can also be found in mid-19th-century photographs of Yosemite that depict a far less densely forested landscape with vast expanses of open meadow and grassland. Contemporary science empirically supports that visual impression, showing how the decreased rhythm and occurrence of fire, through suppression, is correlated with the spread of woody plants that thickened the forests and colonized the mountain meadows.[21] In turn, this shift in plant communities led to much more water uptake by trees, and less runoff and storage in meadow soils.[22] And with the increasing accumulation of all that vegetative biomass, it eventually became impossible to prevent fire events from taking hold, even with costly *firefighting* efforts still in full force. Unlike Yosemite's pre-colonial fires, these higher intensity burns do not just thin out the forest, consume underbrush, and rejuvenate the meadows. Rather, they burn so hot and with such ferocity that they incinerate nearly everything they encounter, leaving behind few, if any trees. They burn subterranean tree roots and cook the soil, leaving behind lifeless, crusty, erodible dirt. The transformation caused by such fire events is akin to a throwback to primary succession; so much so that what these immolated ecosystems formerly were – such as a ponderosa pine forest – might not ever return; instead converting to grassland or scrubland indefinitely.

And Yosemite is not unique. A vast number of landscapes have excess <u>fuel</u> to ignite due to decades of fire suppression techniques. This is the paradox of fire suppression: "the more we try to control burning, the worse the burning that results."[23] In the U.S., fire suppression officially began in the late 1800s, when Army patrols were enlisted to put out fires in Yellowstone, Sequoia, General Grant, and Yosemite National Parks. Then, in 1935, came the U.S. Forest Service's *10 am rule*, which stipulated that all wildland fires needed to be suppressed by 10 am the following day. Over the next decade, firefighting crews were deployed to public lands across the country and thousands of lookout towers were constructed to help identify fires before they got too large to extinguish. In 1944, the U.S. Forest Service enlisted Smokey Bear to promote the idea that fires were primarily accidental and destructive, a misguided public education campaign that was quite successful. But in the late 1960s and 1970s, based on problems observed, policies began to change from wildland fire "control" to management, with the understanding that fire is a vital and necessary landscape process.[24] <u>Fire regime</u> refers to the pattern, frequency, and intensity of wildfires that prevail in an area over time. Fire suppression changes the cadence of fire

Figure 1.3 (*Right*) The cooked soil around a subterranean plant root burned by the LNU Lightning Complex fires of 2020.

regimes, making them longer and less frequent in occurrence. A significant change in fire regime can change a landscape, and if the cadence is too fast or too slow, it can lead to migratory, qualitative changes in what a landscape is, its ecological relations, and how it functions.

But it is not just fire suppression that is the problem. Suppression policies have eased in varying degrees in Yosemite and the Western U.S. over the past 50 years, but combined with trying to work our way out of that time trap, there is also accelerated climate change to contend with. As earth's climate continues to get hotter and destabilize at accelerating rates, it exacerbates risk for areas that are already fire-prone, especially those with a Mediterranean-type climate, like California's. Mediterranean regions tend to have a distinct wet season, when plants sprout and grow, followed by a long dry season, when the vegetation desiccates, dies back, and becomes kindling. With climate change, these regions are experiencing longer and hotter periods of drought, lower levels of humidity and soil moisture, and longer fire seasons that are changing regional plant phenology. All of these factors are further elevating the risk of high intensity fires and risks of fires occurring so frequently that they also cause wholesale ecosystem shifts, such as from forest to shrubland.[25]

A third factor concerns land use and land use changes, such as within the wildland-urban interface (WUI). These are the diverse, liminal spaces between wildland and the built environment. They consist of dispersed housing developments and other built structures and land uses mixed together with remnants of forests, chaparral, and grasslands. In some parts of the world, people are leaving these edge zones to live in urban centers, rendering them feral and derelict, while in others development is aggressively encroaching outward from cities into formerly undeveloped land. In the U.S., for example, the liminal area between wildland and the built environment has grown tremendously over the last few decades, making it one of the fastest growing land use type in the lower 48 states. The presence of humans in this fire-vulnerable landscape is elevating the risk of human-ignited burns. In fact, many of the largest fires in recent years can be attributed to human behavior, be it from vehicle-related sparks, campfires, or downed power lines.[26] The sheer diversity and extent of landscapes emerging and burning in these hybrid zones is hard to overstate.

Just these three factors – changes in fire regime, climate, and land use – in combination, can foster innumerably variable situations for fire events. Interactions among them tend to create aggregating and cascading effects, like the massively-scaled, high-intensity burns happening throughout the Sierra Nevada Mountains and Yosemite, and the deadly Camp Fire that burned through the foothill town of Paradise, CA in 2018, displacing many vulnerable people with limited resources to relocate or rebuild.[27] Climate change tends to accelerate fire regimes (shorter time spans between fires), which can be further accelerated by human use and accidental ignitions, causing migration of woody chaparral to grasslands, which is happening widely in Southern California near sprawling urbanized areas. Added to these factors, are the varying conditions present at the moment when fire takes hold, such as weather, air temperature, dryness, and wind speed and direction. And each distinct type of ecosystem, such as chaparral, conifer forest, riparian floodplain, and grasslands, respond to, and are differently adapted to fire. All these influences act and come together in the immersive, place-specific medium of landscapes.

## Ontologies of Fire

We live in the Central Valley town of Davis, where we are often enveloped in smoky air from wildfires for weeks at a time. Be it spring, summer, or fall, we may see white, papery flakes drifting down from the sky, dusting the ground with ash, and making the sun appear as an ominous red ball in a thick haze. This smoke often contains toxic micro particles from housing developments and other artifacts that have been incinerated in the fires, and air like this can impact much of the state, and the states to the east of it. These fires are feral, having escaped efforts to eradicate and tame them.[28] These pyric events are transforming landscapes at multiple scales – impacting plant communities, animals, air, water, and soils. They make their own weather systems and literally create new kinds of ground in their wake. They are typically larger, hotter, and more destructive, due to how we, as settlers, helped to engender them.

These novel fire conditions are not unique to the American West, as the risk and damages of wildland fires is growing around the world.[29] From catastrophic bushfires in Australia, to severe forest fires in France, to fynbos fires in South Africa, innumerable communities and landscapes are being impacted. Today, this pyrogeography significantly affects every continent except Antarctica. As fire historian Stephen Pyne has argued, this is an age of fire marked by a shift out of the Pleistocene's ice ages and into the hot, combustion-dominated Pyrocene.[30] This geological epoch began nearly 12,000 years ago, when humans harnessed fire and began using it in expanding ways; from a small wood fire to keep warm, to the intentional burning of fields and forests, to the massively-scaled global mining and burning of fossil fuels, which is the literal consumptive burning of another era within our own. In the same way that the Pleistocene's ice ages led to mass extinctions, abrupt changes in sea levels and climate, and massive shifts in the distribution and assemblies of biota, the same type of accelerated changes are happening now, but are happening due to the heating of the planet at unprecedented rates.[31] This is a human-led, coevolutionary change, driven by human desires and design schemes and their impacts. And unfortunately, we are discovering that increased burning and global warming begets more burning. The Pyrocene is only gaining in force, and the changing nature of wildfires is just one piece of this global transition.

For us, Pyne's conception of the Pyrocene provides entry points for understanding the current human-dominated era, calling particular attention to the ways humans have used fire and combustion to alter the earth, deliberately and inadvertently. We think it also provokes more basic ontological questions of how fire is understood and how it is approached. As we observed in Yosemite, colonists considered fire to be destructive, primitive, and bad, and thus did all they could to eradicate it from the U.S. Western landscapes, believing this was progressive stewardship. Their understanding of fire was the opposite of indigenous Californian people that long preceded them. Turtle Island's First Nations understood fire as valuable and consequential, and utilized it far more heavily than any other stewardship practices to sculpt these wildlands.[32] Many of us tend to think of that historic usage as an instrumental *tool*, or *technology*. But that too is a situated understanding based on Western norms. For First Nations, fire was likely thought and felt in very different ways from this and was woven into cosmologies premised on animacy and reciprocity, rather than human exceptionalism and ideologies of extractivism.[33]

Today we speak of "bad" fire or "good" fire, based on the sets of contextual factors that determine its behavior. There is also prescribed, cultural, and suppressed fire. And what does it really *mean* to "fight" fire – symbolically, culturally, materially, and physically? To many ecologists, fire is considered a keystone process in forests and other biomes.[34] It has also been likened to an abiotic herbivore that consumes and physically digests landscape vegetation in binge-like fashion.[35] Yet, it also acts very much like a decomposer. But in the Western scientific world view, fire (and landscapes) is not typically considered to be alive, sacred, or mythical. Rather fire is dealt with instrumentally.

Stephen Pyne observed that although our contemporary societies have all sorts of research and academic departments dedicated to the study of air, water, and soil, the only departments we have had for fire are those charged with putting them out.[36] In post-colonial and post-natural worlds, fire is lacking robust research domains of its own. And where fire is researched, it tends to fall into three ontological paradigms of what fire is, as either physical, biological, or cultural phenomena.[37] The physical paradigm understands fire as primarily a chemical reaction shaped by physical, climatic, and chemical circumstances, and is materially focused on the elemental and reactive processes of combustion. The biological paradigm approaches fire as fundamentally life-based and dependent on the growth and development of biological assemblies of organisms in landscapes, primarily plants. The cultural paradigm encounters fire as largely a social construction, emphasizing human doings and agency in determining the behavior of fire. With respect to perceived problems with fire, each paradigm tends to look to its particular focus for potential explanations and solutions.[38]

But fire is almost always an amalgamation of physical, biological, and cultural factors, and perhaps the best way to see that integration, is to understand fire as a contextual, time-based event. Fire "synthesizes its surroundings," and "its very character is to interact and integrate."[39] Fire is an event that appears "patchily in space and time."[40] And more specific to landscape architecture and planning, fire emerges and takes its form from the landscape in which it occurs.[41]

Fire moves from virtual potential to actualized material process in particular ways, based entirely on the dynamic composition of landscapes. Fire doesn't slowly develop, rather it comes into existence in a flash at a particular moment in time, such as from a lightning strike or a carelessly discarded cigarette butt. But all that comes before it ignites, fully conditions what it will become. After ignition, the flames set about transforming the materiality of the landscape, based on interaction with the specific conditions it finds itself in at that specific moment. In the same way that none of us get to choose our parents, fire doesn't get to choose its landscapes. In both cases, we develop and evolve from those formative settings and potentials, whatever they may be. This is important to consider when labeling fire as "good" or "bad," as the characterization is more about the situation, historical context, and timing, rather than about fire itself.

In this book, we understand and research fire as a contextual landscape event. Fire is entirely dependent on landscape conditions to become what it may, while also being transformative of the conditions from which it emerges. Wildfire is a reciprocal, or coevolutionary phenomena within the dynamic landscapes in which it happens, spanning physical, biological, and cultural domains. As a place-based experience, wildfire is as complex as its spatio-temporal setting.

With this co-creative understanding of fire and landscapes, we explore the changing behavior and effects of fire and illustrate ways of designing with it.

## Fire-Altered Landscapes

If the behavior of wildfire is changing, then the landscapes in which it occurs must be changing too. Historical definitions of landscape, in colonial and Western traditions, have foregrounded stability and passivity as traits, which led to sensibilities of being able to instrumentalize and control them. But within nearly all contemporary disciplines that foreground the medium of landscape – including landscape architecture – this definition has moved to one of dynamism, elasticity, and indeterminacy. There is a broad, empirical understanding that landscapes actively change, and do so unavoidably. We understand landscapes to be a diverse medium marked by migration, or patterned movement across space and time. A landscape can be said to migrate when its unique and contingent assembly of components – the materials, entities, and actors that define it – shifts such that, over time, a new assembly forms. In this way, qualitatively different landscapes can and do manifest upon a single geographic terrain.[42]

Landscapes never stop changing, no matter how hard humans may try to fix them in place. Rather, flows and processes within them are subject to temporal distortion; to variable slowing and acceleration as part of design intention.[43] Landscapes respond to imposed constraints like these by moving *differently*, often mutating into surprising and undesirable manifestations. Fire is but one such actor in these multiply-authored assemblies – a particularly transformative and fast acting one. In this book, we speak of landscapes where humans have significantly or radically changed their qualities by changing how fire occurs within them, as fire-altered landscapes. These landscapes are the focus of our book, and we locate most of them in two general realms: (1) dedicated land reserves (like national parks, forestry reserves, extraction zones, and wildlands) as well as other forms of undeveloped or minimally developed land, and (2) the diverse territory of the WUI.

Yosemite is an example of the land reserve typology. As we've observed, Yosemite and most of California's Sierra Nevada mountains became fire-altered landscapes as a result of decades of fire suppression, thus changing it, rather than preserving it. This paradox is a common one encountered in 20th-century ecological reserves, national parks, conservation areas, and landscape preservation efforts generally. Many of these efforts failed, spectacularly, by suppressing the very landscape dynamics that sustain landscape structure over time. This legacy is proving to be an extremely difficult and destructive one to rectify, and in response to these mistakes, conservation strategies have been evolving in ways that are more integrative of change and seek to work with, rather than against, dynamic, landscape processes.[44] Today, Yosemite's park staff are engaged in a range of proactive land-fire stewardship efforts to try to pull out of the fire suppression trap, including managing wildfires rather than suppressing them, prescribed burns, and mechanical thinning and removal of crowded trees within the state's most famous valley, which doesn't happen without controversy and a lawsuit.[45]

The other broad domain of fire-altered landscape is the WUI briefly mentioned above, where feral flames often abound. This peripheral zone has been labeled many things in the fields of landscape architecture and planning – from "exurban"[46] to "rural-urban fringe"[47]

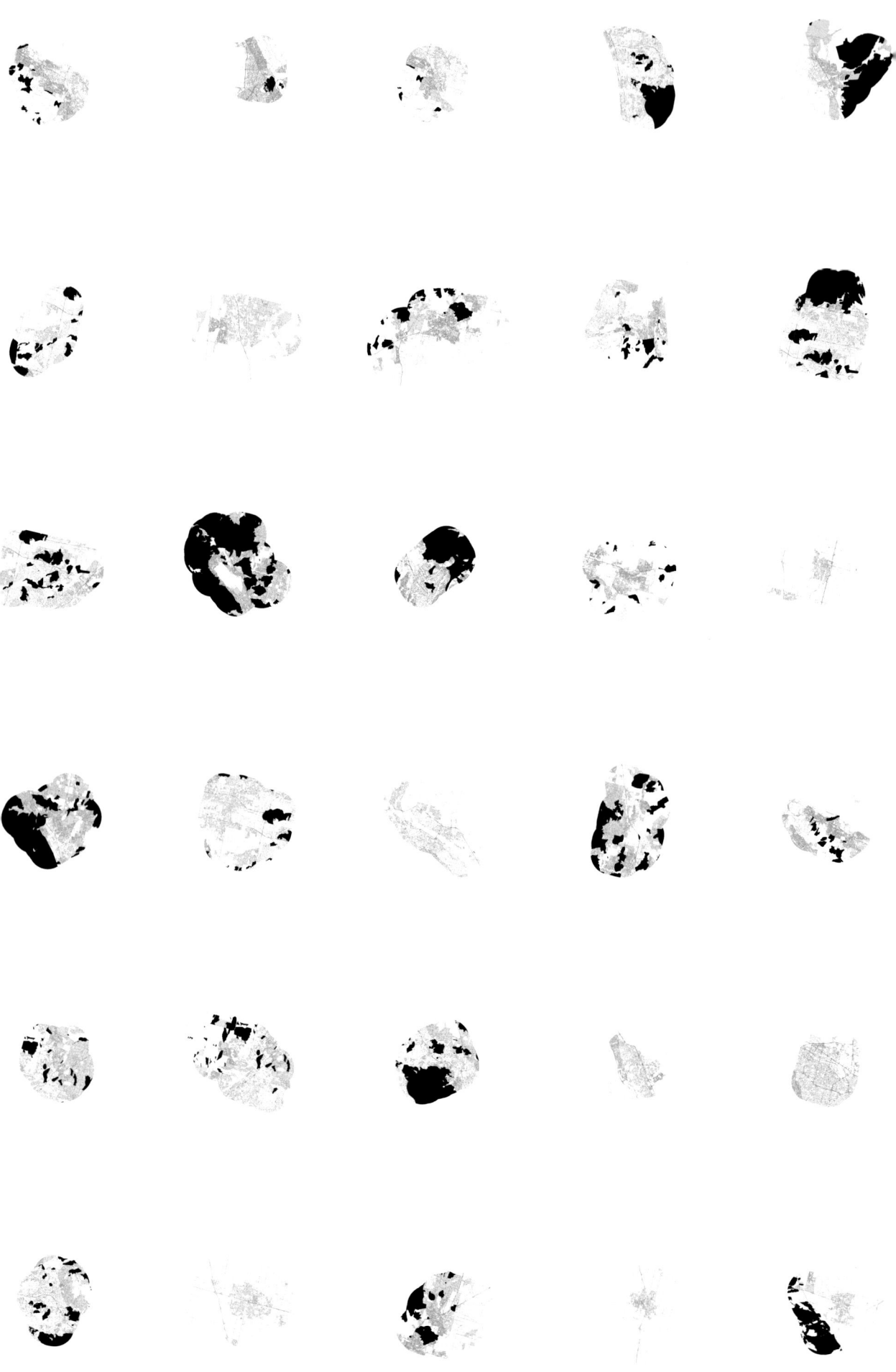

to "borderlands"[48] – with the understanding that wildland and the built environment are not binary but are, in fact, hybridic, diverse, and intertwined.

To better understand this space, two terms often put at odds with one another need to be refashioned: "wilderness" and "city." For Europeans and American settlers, the concept of wilderness has morphed in the past 250 years, from instilling fear to instilling awe. Up until the end of the 19th century, a common belief was that wilderness had no real value to humanity. Rather, it was a source of darkness and terror. Then came Muir, Olmsted, and others, who flipped this narrative and advocated for the preservation of wilderness as a place of restorative escape; as pristine and external to human agency.[49] This take on wilderness, which was similar to Olmsted's, was codified in the U.S. Wilderness Act of 1964:

> A wilderness, in contrast with those areas where man and his works dominate the landscape, is hereby recognized as an area where the earth and its community of life are untrammeled by man, where man himself is a visitor who does not remain. An area of wilderness is further defined to mean in this Act an area of undeveloped Federal land retaining its primeval character and influence, without permanent improvements or human habitation, which is protected and managed so as to preserve its natural conditions and which (1) generally appears to have been affected primarily by the forces of nature, with the imprint of man's work substantially unnoticeable; (2) has outstanding opportunities for solitude or a primitive and unconfined type of recreation; (3) has at least five thousand acres of land or is of sufficient size as to make practicable its preservation and use in an unimpaired condition; and (4) may also contain ecological, geological, or other features of scientific, educational, scenic, or historical value.[50]

But there is increasing understanding that wilderness, as "other," is untenable, and is actually a "product of civilization."[51] The imposed European concept of wilderness was likely unintelligible for North America's indigenous nations, particularly in California, where people had extensively tended to and shaped most of its landscapes.[52] The adept landscape architecture of pre-colonial California was so extensive and so seamlessly integrated, that most settlers could not see or detect it, or else didn't want to, and thus fictitiously called it wilderness. Through time, these definitions of wilderness and that of the Wilderness Act have, in some ways, made their definition real by enforcing their policies and managerial separations of humans and "nature."

In parallel, the concept of the city has also radically changed over past centuries. Historically, cities were discrete and coherent entities, with clearly defined boundaries separating civilization from what was outside of its borders. Today, though, it is quite obvious that the *urban* and processes of urbanization systemically extend far beyond city edges, touching nearly every part of the landscape and deeply penetrating into the assumed territories of wilderness.[53]

The intermingling of wildland and the built environment has become a focus of much research interest over the past few decades. As a term, WUI first emerged in the late 1980s in the United States but was not formally incorporated into federal management practices until 2000.[54] Today, while the term has been adopted across many fire-prone parts of the world, there is no universally accepted definition for it. It has been described

Figure 1.4 (*Previous*) Spatial comparison of California's fastest growing cities, showing wildland-urban interface areas (gray) and historical fire perimeters (black).

qualitatively, in very general terms, as a space where "humans and their development meet or intermix with wildland _fuel_."[55] It has also been described quantitatively, using housing units per acre and vegetation coverage percentages as criteria.[56] In examining the maps that accompany these definitions, it becomes clear that the WUI, as a condition, now extends to many seemingly unexpected places, including Olmsted's granite cathedral of the West, Yosemite.[57]

Like many other liminal landscapes, this motley interface is both a place of vulnerability and opportunity. From one perspective, it provides desired access to open space, with scenic views and cleaner air. Additionally, these edge zones often provide a more affordable living option compared to high rents and property values found closer to city centers, though in places like California, this price gap is shrinking. Development in these edges accelerates habitat loss and fragmentation and supports unintended introduction of exotic species and diseases, while simultaneously promoting vehicular-dependent development. And situated on the frontline of the _Pyrocene_, these edge zones are highly susceptible to fire, and housing adds more fuel to burn, as well as toxins to the smoke that many of us now breathe.

## Designing with Fire

Feral wildlands and the WUI are places where design can make a profound impact and where designers and landscape architects can effectively work with others to guide fire in the deepening age of the _Pyrocene_. But to do so, it must be approached integratively and inclusively, rather than reductively and simplistically.[58] With this in mind, the following chapters have been developed using four principles: design across scales, embrace accelerated change, co-design, and advance transdisciplinarity practices.

The first principle – _design across scales_ – is expressed through a wide range of design case studies, from regional down to the cellular, and from decades down to moments. Scale – as the spatial or temporal extent at which something is examined – is a purely conceptual design tool; a single point of reference for investigation that does not physically exist in the world. To focus on a single scale, or a limited, codified set of scales, limits what can be understood in design research. For fire adaptation, multi-scalar thinking and designing is far more informative and necessary. Different forces and processes come to the foreground at different extents of observation. Thus, we fluidly jump scales within and across project case studies, and have intentionally avoided ordering them by size of intervention, given that those extents are hard to cleanly determine, such as comparing a modular intervention at the small scale of a residence (like "defensible space"), to large-scale land management. Categorizing these by scale would be overly reductive of the many other dependent factors, as well as the qualitative attributes of the approaches. The content of this book aims to expand the design toolbox to provide a diverse and variably scaled set of design techniques that can be collectively critiqued, modified, and tested to enact change.

_Embrace accelerated change_ emphasizes the challenges and conditions presented by global warming, as well as increasing social and economic disparities driven by colonial and capitalist policies and agendas. This principle integrates increasing rates of change as a globalized given, and landscape as the elastic and responsive medium in which these altered rhythms are enacted and given form. This book promotes process-based, adaptive strategies that perform their work over time, across a chaordic string of events and situations and favors the design of relations over objects. It promotes iterative design proposals that are committed to adaptive stewardship and care; favoring strategies that revisit past actions to assess and adjust formative design pathways.

Designing with fire is far more than a scientific or technical challenge or domain. How landscape fire happens or doesn't happen is also fundamentally social and political concerns. Neither suppression nor science alone adequately addresses those concerns. Thus, our third principle – *codesign* – focuses on fostering greater inclusivity and equity in design processes. Co-design entails bringing stakeholders and publics into design and planning processes who are often excluded from it and who might be impacted by such efforts. It entails bringing people into the process as early as possible. Our research has sought to highlight case studies and efforts to democratize and decolonize fire design and management, and to foster environmental justice and collective social innovation, including better understanding and respect for traditional ecological knowledge (TEK) and indigenous sovereignty.

In definition, transdisciplinary entails creative collaboration across disciplines in a manner that meaningfully integrates publics and stakeholders in that process. This book seeks to *advance transdisciplinarity practices* in planning and design by documenting integrative approaches to wildland fire adaptation that transcend current institutional, organizational, and research-related silos.[59] Given the complexity and global scale of fire stewardship, it is clear that no single discipline or organization has all the answers to making things better, and there is an urgent need to develop effective ways to design across disciplines with the people who will be affected by such efforts. In this kind of collaborative work, we position designers as integrators and catalysts of these design efforts, testing the range and efficacy of such roles.

## Approaches for Evolving and Adapting

This book is focused on fire-altered landscapes, with the sense that we are on a cultural, environmental, and economic threshold in living with fire. By presenting a framework for designing in and with these landscapes, our hope is to inspire and challenge landscape architects, planners, policy makers and other disciplines to reconceptualize and reimagine what pyro futures we might create. Rather than serving as a book of neatly packaged solutions, it includes a compendium of techniques to be considered, evaluated, and tested. More than anything else, the following chapters aim to elevate the conversation about the integrative role landscape architects might play in designing in a time of fire.

In terms of the structure of the book, after this introduction, the next chapter offers a deeper description of fire-altered landscapes, discussing where and what they are, and the challenges they present. Chapter three provides a glossary of fire-related terms, which appear as underlined text throughout the book. After the glossary, the next three chapters are a curated collection of design case studies from around the world, both built

Figure 1.5 (*Right*)
The 27 case studies featured in this book, organized into three general approaches.

**Resistance**

01 Fire Fighting
02 Foil Wrapping
03 Site Hardening
04 Home Bunkering
05 Firebreak Cutting
06 Ridge Clearing
07 Donut Extracting
08 Defensive Spacing
09 Earth Shifting
10 Ring Tending
11 Agricultural Stripping

**Co-Creation**

12 Patchy Planting
13 Ember Trapping
14 Infrastructure Shadowing
15 Cyborg Landscaping
16 Selective Thinning
17 Invasive Hacking
18 Refugia Expanding
19 Block Burning
20 Firestick Farming
21 Fire Lighting
22 Fire Flocking

**Retreat**

23 Development Limiting
24 Construction Halting
25 Incentivized Relocating
26 Wholesale Moving
27 Fire Surrendering

and speculative. Lastly, the final chapter closes with speculation on what pyro-based scenarios and futures we might inhabit, based on the types of landscape adaptation techniques that are, or are not enacted in coming years.

As authors, our experience is mostly grounded in the fire-altered landscapes of California, where the severity and magnitude of fire issues is a driving motivation for our writing this book, and where we are actively researching and testing landscape-based techniques to more skillfully coevolve with fire. Thus, we present examples from within California as well as similar Mediterranean climate landscapes around the world for the exchange of ideas and conversations they can foster, particularly when so little of such work has been assembled for designers.

The case study chapters have been organized into three overarching approaches to fire that we have observed in our research: resistance, co-creation, and retreat, which we describe below.

## *Resistance*

These are approaches that tend to take a stand against the creative and transformative propensities of fire and forces of landscape change. To resist, in this context, is to try to control, or to maintain a desired status quo of landscapes and how they are used and occupied. Resistance is often characterized by the continuation and augmentation of established design, policy, and management techniques that are increasingly challenged by the effects of past efforts (whether intended or not), as well as new and emergent challenges brought about by a rapidly changing climate and pervasive landscape alteration.

Generally, this approach seeks to preserve and maintain human activities, infrastructures, values, development patterns, and investments as they are, thus resisting or potentially delaying more systematic change and socio-eco-technical adaptation.[60] These approaches often emphasize immediate social concerns – such as political constituencies and economic gains – which can come at the expense of longer-term planning, and thus through resistance, increase risks and vulnerabilities over longer time horizons.

Yet, there are many pressing and compelling reasons for implementing resistance approaches, such as protecting critical infrastructures that society needs to function in its current configuration, and protecting existing communities. Human adaptation and socio-technical evolution to changing conditions and values usually takes time, and thus there is a need for transitional landscapes and design practices to be implemented until more significant adaptation may occur, which often entails deeper and more difficult transitions in ideologies, politics, and the power structures of resistance. However, such efforts can also impede or slow the development of more adaptive measures at a time when they are urgently needed to avoid more harmful impacts or the inability to adapt due to delayed action, which is why some of these strategies may be considered maladaptive in the context of global accelerated change.[61] Thus, each technique that fits with this category should be assessed according to its particular regional context (including governance structures, economy, equity, and environmental justice, physical geography), as well as larger global trends.

## Co-creation

These are approaches to fire and fire stewardship that intentionally embrace and utilize landscape forces, while also trying to intentionally and creatively guide them. A co-creative approach is one of give and take and feedback between people and landscapes, as exemplified by pre-colonial land stewardship in North America. It is one that acknowledges both the agency and generative capacity of fire, and seeks to work with it. It's an approach that understands that landscapes cannot be controlled, but can be adaptively stewarded and collaborated with. Co-creation implies a lack of singular and distinct authorship in these techniques, as agency is broadly shared and distributed across landscape assemblies, people, and climates.

We understand co-creative approaches as spanning a loose and diverse middle ground between the poles of resist and retreat. It works in the realm of active care, design, effort, learning, and change.

## Retreat

In contrast to resistance approaches, retreat is the "letting go" of perceived control or dominion of a landscape; of giving it over to itself to evolve and become. From a design perspective, the most important question about retreat is whether, when it occurs, if it is intentional and designed (like *managed* or *strategic* retreat) or forced and catastrophic, due to lack of preparation or action. Either can happen. Societies may end up retreating from some landscapes, in terms of management or living in them, simply because it isn't feasible or safe anymore. When retreat is intentional and strategic, it can be defined as minimizing management of a landscape to let it become what it may, and the coordinated movement of people, buildings, communities, and infrastructures away from areas and landscapes of perceived high risk. Retreat is most commonly understood and discussed in relation to flood risk and sea level rise, but it is applicable across an increasing range of environmental risks, such as drought, excessive heat, toxicity, and wildfire.

In most contexts, retreat is the very opposite of maintaining the cultural and ecological status quo. Accordingly, retreat is typically considered "politically toxic," "infeasible," and even "impossible,"[62] particularly in the neoliberal context of the U.S., where growth and extractive occupation of landscapes is a deeply embedded economic ideology. But as Liz Koslov has described:

> *Retreat is a powerful and controversial concept whose cultural and political significance will grow as the planet warms and the seas rise, and it is already a valuable and necessary addition to the language of climate change adaptation. Without taking retreat seriously as a concept, strategy, and existing practice, meaningful conversation and action around climate change adaptation will continue to prove illusory. Understanding community-organized relocation efforts as forms of retreat unifies this emerging practice with other social movements and political projects that seek more sustainable ways of settling on earth.*[63]

As we progress into the future, and extreme droughts become more common and fire seasons continue to extend or become year round, retreat will shift from far-fetched to necessary. Already, in states like California, the scale of the wildfire issue – exacerbated by resistance strategies of fire suppression – is far beyond the scale at which it can be feasibly "managed." As more communities in the WUI continue to burn and reburn at greater frequencies, insurance plans will become unfeasible. In the long term, we understand retreat – in relation to wildfire – to be inevitable and unavoidable. The real question is whether retreat will be designed and done with foresight, or if it will happen catastrophically and inequitably, through lack of such efforts. Either way, adaptation by retreat is poised to emerge as a prominent adaptation approach as we advance deeper into a rapidly changing climate.

Retreat is not without significant design challenges, including how its rewildings and relocations are approached in an equitable and participatory manner, and in deciding what is let go. Retreat can occur across a range of bottom-up to top-down strategies. Given its departure from the status quo, retreat will likely require shifts in cultural attitudes, economic regimes, and ideologies if it is to occur at the scale at which it will be needed. In short, retreat will require broader social innovation and adaptation, where a range of more "sustainable ways of settling on earth," such as degrowth – the anathema to capitalism – will be in greater circulation and debate.

Additionally, many landscapes have been so radically altered, that if we, as occupiers, stop caring for them, they will not act "naturally," but rather quite ferally. Due to a plethora of anthropogenic alterations in the cycling of matter, species, and climate, highly undesirable results can occur when humans pack up and leave a place. Thus, many design questions with retreat focus on *how* to step away, and *to what degree* it is possible to do so, since in so many places of the U.S., landscapes were sculpted and cared for thousands of years prior to colonial appropriation. To remove human stewardship from them would be as novel as many of these ecosystems now are.

The boundaries and distinctions between these three adaptive approaches to fire are porous, rather than distinct. Overall, our hope is that the case studies serve as an entry point for further design imagination and research, both for now and feeling into the future.

# Notes

1   Robin Kimmerer et al., "Maintaining the Mosaic: The Role of Indigenous Burning in Land Management," *Journal of Forestry – Washington* 99, no.11 (2001): 36–41.

2   Rebecca Solnit, *Savage Dreams: A Journey into the Hidden Wars of the American West* (Oakland: University of California Press, 2014): 308.

3   Frederick Law Olmsted et al., "The Yosemite Valley and the Mariposa Big Trees: A Preliminary Report (1865)," *Landscape Architecture Magazine* 45, no.1 (October 1952): 12–25.

4   Ibid.

5   Ibid.

6   Ibid.

7   Ibid.

8   Ibid.

9   Frederick Law Olmsted, "Governmental Preservation of Natural Scenery," in *The Papers of Frederick Law Olmsted: The Early Boston Years, 1882–1890*, ed. Ethan Carr, Amanda Gagel, and Michael Shapiro (Baltimore: Johns Hopkins University Press, 2013): 33.

10  Charles Beveridge, "Olmsted and Yosemite," *SiteLINES* 5, no.1 (2009): 6–8.

11  Mark Spence, "Dispossessing the Wilderness: Yosemite Indians and the National Park Ideal, 1864–1930," *Pacific Historical Review* 65, no.1 (1996): 27–59.

12  M. Kat Anderson, *Tending the Wild: Native American Knowledge and the Management of California's Natural Resources* (Oakland: University of California Press, 2013), 3.

13  Lina Gassaway, "Native American Fire Patterns in Yosemite Valley: A Cross-Disciplinary Study," *Proceedings of the 23rd Tall Timbers Fire Ecology Conference: Fire in Grassland and Shrubland Ecosystems* 23 (2007): 29–39.

14  Elizabeth Godfrey, *Yosemite Indians: Yesterday and Today* (Yosemite: Yosemite Natural History Association, 1941).

15  Frederick Law Olmsted, and Laura Wood Roper. "The Yosemite Valley and the Mariposa Big Trees: A Preliminary Report (1865)." *Landscape Architecture Magazine* 45, no.1 (October 1952): 12–25.

16  William M. Denevan, "The 'Pristine Myth' Revisited," *Geographical Review* 101, no.4 (October 2011): 576–591.

17  Cronon, William, "The Trouble with Wilderness: Or, Getting Back to the Wrong Nature," *Environmental History* 1, no.1 (1996): 7–28.

18  Wendy Harding, "Frederick Law Olmsted's Failed Encounter with Yosemite and the Invention of a Proto-Environmentalist," *Ecozon* 5, no.1 (2014): 123–135.

19  Ibid.

20  Ibid.

21  Van Kane et al., "Landscape-Scale Effects of Fire Severity on Mixed-Conifer and Red Fir Forest Structure in Yosemite National Park," *Forest Ecology and Management* 287 (2013): 17–31.

22  Scott Stephens et al., "Fire, Water, and Biodiversity in the Sierra Nevada: A Possible Triple Win," *Environmental Research Communications* 3, no.8 (2021): 1–10.

23  Stephen Pyne, *Fire: A Brief History* (Seattle: University of Washington Press, 2001): 193.

24  Jan W. van Wagtendonk, "The History and Evolution of Wildland Fire Use," *Fire Ecology* 3 (2007): 3–17.

25  Yongqiang Liu et al., "Trends in Global Wildfire Potential in a Changing Climate," *Forest Ecology and Management* 259, no.4 (2010): 685–697.

26  Volker Radeloff et al., "Rapid Growth of the US Wildland-Urban Interface Raises Wildfire Risk," *PNAS* 15, no.13 (2018): 3314–3319.

27  Justine Calma, "What Losing Paradise Tells Us About Today's Blazes," *The Verge*, accessed November 11, 2021, https://www.theverge.com/22652695/paradise-california-wildfires-fire-season-dixie-caldor

28  A. Park Williams et al., "Observed Impacts of Anthropogenic Climate Change on Wildfire in California," *Earth's Future* 7, no.8 (2019): 892–910.

29  Matthew W. Jones et al., "Climate Change Increases the Risk of Wildfires," *ScienceBrief Review* (2020): 1–3.

30  Stephen Pyne, *The Pyrocene: How We Created an Age of Fire, and What Happens Next* (Oakland: University of California Press, 2021).

31  Stephen Pyne, "The Planet is Burning," Aeon, accessed November 11, 2021, https://aeon.co/essays/the-planet-is-burning-around-us-is-it-time-to-declare-the-pyrocene

32  Anderson, *Tending the Wild*.

33  Robin Kimmerer, "Learning the Grammar of Animacy," *Anthropology of Consciousness* 28, no.2 (Fall 2017): 128–134.

34  Kane et al., "Landscape-Scale," 17–31.

35  William J. Bond et al., "Fire as a Global 'Herbivore': The Ecology and Evolution of Flammable Ecosystems," *Trends in Ecology & Evolution* 20, no.7 (2005): 387–394.

36  Pyne, *Fire*, 187–201.

37  Ibid.

38  Ibid.

39  Pyne, *The Pyrocene*, 9–12.

40  Pyne, *The Pyrocene*, 15.

41  Pyne, *The Pyrocene*, 14.

42   Brett Milligan, "Landscape Migration: Environmental Design in the Anthropocene," *Places Journal* (June 2015), accessed November 11, 2021, https://placesjournal.org/article/landscape-migration/#0

43   Brett Milligan, "Accelerated and Decelerated Landscapes," *Places Journal*, February 2022, accessed October 29, 2022, https://doi.org/10.22269/220208

44   For example, see: Jedediah Purdy, *After Nature* (Cambridge: Harvard University Press, 2015), Jamie Lorimer, *Wildlife in the Anthropocene: Conservation after Nature* (Minneapolis: University of Minnesota Press, 2015), and Eric Higgs, *Nature by Design: People, Natural Process, and Ecological Restoration* (Cambridge: MIT Press, 2003).

45   Thomas Fuller and Livia Albeck-Ripka, "At Yosemite, a Preservation Plan That Calls for Chain Saws," *The New York Times*, https://www.nytimes.com/2022/07/27/us/yosemite-fires-cut-and-burn.html

46   Peter Calthorpe, *The Next American Metropolis: Ecologies, Communities, and the American Dream* (Hudson: Princeton Architectural Press, 1993).

47   Robin J. Pryor, "Defining the Rural-Urban Fringe," *Social Forces* 47, no.2 (1968): 202–215.

48   John Stilgoe, *Borderland: Origins of the American Suburb, 1820–1939* (New Haven: Yale University Press, 2012).

49   Cronon, "The Trouble With."

50   "Wilderness Act, 1964," accessed November 11, 2021, https://www.nps.gov/parkhistory/online_books/anps/anps_6b.htm#:~:text=(c)%20A%20wilderness%2C%20in,visitor%20who%20does%20not%20remain.

51   William Cronon, "The Trouble With."

52   Anderson, *Tending the Wild*.

53   For example, see: Henri Lefebvre, *The Urban Revolution* (Minneapolis: University of Minnesota Press, 2003). And Neil Brenner, *Implosions/Explosions: Towards a Study of Planetary Urbanization* (Berlin: JOVIS, 2014).

54   William Sommers, "The Emergence of the Wildland-Urban Interface Concept," *Forest History Today* (Fall 2008): 12–18.

55   Susan Stein et al. "Wildfire, Wildlands, and People: Understanding and Preparing for Wildfire in the Wildland-Urban Interface," USDA Forest Service, accessed September 9, 2020, https://www.fs.fed.us/openspace/fote/reports/GTR-299.pdf

56   Ibid.

57   Volker et al., "Rapid Growth," 3314–3319.

58   Richard Weller et al., "Hotspot Cities: Identifying Peri-Urban Conflict Zones," *Journal of Landscape Architecture* 14, no.1 (2019): 8–19.

59   Alistair Smith et al., "The Science of Firescapes: Achieving Fire-Resilient Communities," *BioScience* 66, no.2 (2016): 130–146.

60   Andrew Cooper et al., "The Adaptation-Resistance Spectrum: A Classification of Contemporary Adaptation Approaches to Climate-Related Coastal Change," *Ocean & Coastal Management* 94 (2014): 90–98.

61   Ibid.

62   Liz Koslov, "The Case for Retreat," *Public Culture* 28, no.2 (2016): 359–387.

63   Ibid.

# Bibliography

Anderson, M. Kat. *Tending the Wild: Native American Knowledge and the Management of California's Natural Resources*. Oakland: University of California Press, 2013.

Beveridge, Charles. "Olmsted and Yosemite." *SiteLINES* 5, no.1 (2009): 6–8.

Bond, William J., and Jon E. Keeley. "Fire as a Global 'Herbivore': The Ecology and Evolution of Flammable Ecosystems." *Trends in Ecology & Evolution* 20, no.7 (2005): 387–394.

Brenner, Neil. *Implosions/Explosions: Towards a Study of Planetary Urbanization*. Berlin: JOVIS, 2014.

Calma, Justine. "What Losing Paradise Tells Us About Today's Blazes." *The Verge*. Accessed November 11, 2021. https://www.theverge.com/22652695/paradise-california-wildfires-fire-season-dixie-caldor

Calthorpe, Peter. *The Next American Metropolis: Ecologies, Communities, and the American Dream*. Hudson: Princeton Architectural Press, 1993.

Cooper, Andrew, and Jeremy Pile. "The Adaptation-Resistance Spectrum: A Classification of Contemporary Adaptation Approaches to Climate-Related Coastal Change." *Ocean & Coastal Management* 94 (2014): 90–98.

Cronon, William. "The Trouble with Wilderness; or, Getting Back to the Wrong Nature." In *Uncommon Ground: Rethinking the Human lace in Nature*. Ed. William Cronon. New York: W. W. Norton & Co., 1995.

Denevan, William M. "The "Pristine Myth Revisited." *Geographical Review* 101, no.4 (October 2011): 576–591.

Fuller, Thomas and Livia Albeck-Ripka. "At Yosemite, a Preservation Plan That Calls for Chain Saws". *The New York Times*, https://www.nytimes.com/2022/07/27/us/yosemite-fires-cut-and-burn.html

Gassaway, Lina. "Native American Fire Patterns in Yosemite Valley: A Cross-Disciplinary Study." *Proceedings of the 23rd Tall Timbers Fire Ecology Conference: Fire in Grassland and Shrubland Ecosystems* 23 (2007): 29–39.

Godfrey, Elizabeth. *Yosemite Indians: Yesterday and Today*. Yosemite: Yosemite Natural History Association, 1941.

Harding, Wendy. "Frederick Law Olmsted's Failed Encounter with Yosemite and the Invention of a Proto-Environmentalist." *Ecozon* 5, no.1 (2014): 123–135.

Higgs, Eric. *Nature by Design: People, Natural Process, and Ecological Restoration*. Cambridge: MIT Press, 2003.

Jones, Matthew W., Adam Smith, Richard Betts, Josep G. Canadell, I. Colin Prentice, and Corinne Le Quere. "Climate Change Increases the Risk of Wildfires." *ScienceBrief Review* (2020): 1–3.

Kane, Van, James A. Lutz, Susan I. Roberts, Douglas F. Smith, Robert J. Mcgaughey, Nicholas Povak, and Matthew Lamar Brooks. "Landscape-Scale Effects of Fire Severity on Mixed-Conifer and Red Fir Forest Structure in Yosemite National Park." *Forest Ecology and Management* 287 (2013): 17–31.

Kimmerer, Robin. "Learning the Grammar of Animacy." *Anthropology of Consciousness* 28, no.2 (Fall 2017): 128–134.

Kimmerer, Robin, and Frank Lake. "Maintaining the Mosaic: The Role of Indigenous Burning in Land Management." *Journal of Forestry – Washington* 99, no.11 (2001): 36–41.

Koslov, Liz. "The Case for Retreat." *Public Culture* 28, no.2 (2016): 359–387.

Lefebvre, Henri. *The Urban Revolution*. Minneapolis: University of Minnesota Press, 2003.

Liu, Yongqiang, John Stanturf, and Scott Goodrick. "Trends in Global Wildfire Potential in a Changing Climate." *Forest Ecology and Management* 259, no.4 (2010): 685–697.

Lorimer, Jamie. *Wildlife in the Anthropocene: Conservation After Nature*. Minneapolis: University of Minnesota Press, 2015.

Milligan, Brett. "Landscape Migration: Environmental Design in the Anthropocene." *Places Journal* (June 2015). Accessed November 11, 2021. https://placesjournal.org/article/landscape-migration/#0

Milligan, Brett. "Accelerated and Decelerated Landscapes." *Places Journal* (February 2022). Accessed October 29 2022. https://doi.org/10.22269/220208

Olmsted, Frederick Law. "Governmental Preservation of Natural Scenery." In *The Papers of Frederick Law Olmsted: The Early Boston Years, 1882–1890*. Ed. Ethan Carr, Amanda Gagel, and Michael Shapiro. Baltimore: Johns Hopkins University Press, 2013.

Olmsted, Frederick Law, and Laura Wood Roper. "The Yosemite Valley and the Mariposa Big Trees: A Preliminary Report (1865)." *Landscape Architecture Magazine* 45, no.1 (October 1952): 12–25.

Pryor, Robin J. "Defining the Rural-Urban Fringe." *Social Forces* 47, no.2 (1968): 202–215.

Purdy, Jedediah. *After Nature*. Cambridge: Harvard University Press, 2015.

Pyne, Stephen. *Fire: A Brief History*. Seattle: University of Washington Press, 2001.

Pyne, Stephen. "The Planet is Burning." Aeon. Accessed November 11, 2021. https://aeon.co/essays/the-planet-is-burning-around-us-is-it-time-to-declare-the-pyrocene

Pyne, Stephen. *The Pyrocene: How We Created an Age of Fire, and What Happens Next*. Oakland: University of California Press, 2021.

Radeloff, Volker, David P. Helmers, H. Anu Kramer, Miranda H. Mockrin, Patricia M. Alexandre, Avi Bar-Massada, Van Butsic, et al. "Rapid Growth of the US Wildland-Urban Interface Raises Wildfire Risk." *PNAS* 15, no.13 (2018): 3314–3319.

Solnit, Rebecca. *Savage Dreams: A Journey into the Hidden Wars of the American West*. Oakland: University of California Press, 2014.

Sommers, William. "The Emergence of the Wildland-Urban Interface Concept." *Forest History Today* (Fall 2008): 12–18.

Smith, Alistair, Crystal A. Kolden, Travis B. Paveglio, Mark A. Cochrane, David M. Bowman, Max A. Moritz, Andrew D. Kliskey, et al. "The Science of Firescapes: Achieving Fire-Resilient Communities." *BioScience* 66, no.2 (2016): 130–146.

Spence, Mark. "Dispossessing the Wilderness: Yosemite Indians and the National Park Ideal, 1864–1930." *Pacific Historical Review* 65, no.1 (1996): 27–59.

Stein, Susan, Sara J. Comas, James P. Menakis, Mary A. Carr, Susan I. Stewart, Helene Cleveland, Lincoln Bramwell, and Volker C. Radelofff. "Wildfire, Wildlands, and People: Understanding and Preparing for Wildfire in the Wildland-Urban Interface." USDA Forest Service. Accessed September 9, 2020. https://www.fs.fed.us/openspace/fote/reports/GTR-299.pdf

Stephens, Scott, Sally Thompson, Gabrielle Boisrame, Brandon M. Collins, Lauren C. Ponisio, Ekaterina Rakhmatulina, Zachary L. Steel, Jens T. Stevens, Jan W. van Wagtendonk, and Kate Wilkin. "Fire, Water, and Biodiversity in the Sierra Nevada: A Possible Triple Win." *Environmental Research Communications* 3, no.8 (2021): 1–10.

Stilgoe, John. *Borderland: Origins of the American Suburb, 1820–1939*. New Haven: Yale University Press, 2012.

Wagtendonk, Jan W. van. "The History and Evolution of Wildland Fire Use." *Fire Ecology* 3 (2007): 3–17.

Weller, Richard, Zuzanna Dozdz, and Sara Padgett Kjaersgaard. "Hotspot Cities: Identifying Peri-Urban Conflict Zones." *Journal of Landscape Architecture* 14, no.1 (2019): 8–19.

"Wilderness Act, 1964." Accessed November 11, 2021. https://www.nps.gov/parkhistory/online_books/anps/anps_6b.htm#:~:text=(c)%20A%20wilderness%2C%20in, visitor%20who%20does%20not%20remain

Williams, A. Park, John T. Abatzoglou, Alexander Gershunov, Janin Guzman-Morales, Daniel A. Bishop, Jennifer K. Balch, and Dennis P. Lettenmaier. "Observed Impacts of Anthropogenic Climate Change on Wildfire in California." *Earth's Future* 7, no.8 (2019), 892–910.

# Landscapes of Fire

Figure 2.1

View of a landscape and highway around Lake Berryessa after the LNU Lightning Complex
Fires of 2020.

# Chapter 2

# Landscapes of Fire

The emerging discipline of <u>pyrogeography</u> considers wildfires not as isolated local events, but as connected events that are part of a larger system. One global geography that is particularly important within this discussion is the Mediterranean-type climate (MTC) zone which has five primary regions scattered across the world: the Cape, Central Chile, the Mediterranean Basin, South and Southwest Australia, and Western North America.[1]

All five of these areas are located between 30 and 40 degrees north and south of the equator and generally sit on the southwestern side of large land masses with high topographic variation.[2] These areas share similar climatic conditions and are generally defined by hot and dry summers and mild and wet winters. With climate change, rainfall in these regions is expected to change and be more variable. In terms of plant communities, they are structurally similar, comprising primarily evergreen shrubs, semi-deciduous scrub, and forests. They are also highly diverse – despite comprising only 5% of the earth's surface, these regions host nearly 20% of the earth's vascular plant species and are considered biodiversity hotspots.[3]

Due to their mild seasons and close proximity to the coast, MTC regions are also highly desirable places to live and are often home to large population centers. As a result, many areas have undergone significant landscape conversions with urban, agricultural, and industrial development replacing and fragmenting native habitat.[4] Significant biodiversity loss is a critical concern for MTC regions – an issue that is further complicated by lagging landscape management practices and complex land ownership and policy patterns.[5] Additionally, fire is an important ecological process in this zone and all five MTC regions have highly predictable wildfire seasons. In recent decades, though, wildfires in the MTC have been exacerbated by climate change and human activity.[6] For example, in some forested areas, decades of human-induced fire suppression have resulted in an accumulation of <u>fuels</u>, increasing the likelihood of large and intense wildfires. And in some shrublands, an increase in <u>anthropogenic fires</u> has resulted in these landscapes burning more frequently than they should.[7] Thus, the MTC zone is a space of tension, with managers having to balance wildfire risk reduction to protect local communities and critical infrastructure while allowing for fire to play its critical ecological role in the landscape.

The MTC zone, is not only a geography of intense vulnerability but also one of unique opportunity, and is the primary focus of this book.

DOI: 10.4324/9781003172956-3

Worcester

Cape Town

☐ Csb- Warm-summer Mediterranean climate

▦ Csa- Hot-summer Mediterranean climate

⊞      100 mi

# A

## *The Cape*

The MTC region of the Cape, located on the southwestern coast of South Africa, includes the cities of Cape Town, Stellenbosch, Worcester, Paarl, and George. Its topography is rugged and diverse with a set of sandstone mountain ranges that run parallel to one another.

From a vegetative perspective, the Cape is extraordinarily diverse partially due to its nutrient poor substrates. It is the smallest and richest of the world's six floristic kingdoms, boasting a high number of species that can only be found in the region. One area of particular interest is a 50–150-mile-wide belt of fire-prone shrublands called fynbos that hugs the coast of the country. Here, fires are needed every 10–15 years to rejuvenate the landscape and maintain the fire-adapted species. The fynbos is also an area of high economic value as plants are often used and exported internationally for building materials, medicine, gardens, and floral design.[8]

While reoccurring fires were a large part of indigenous African culture in the region, they were largely stopped after European colonization in the mid-1600s, and it was not until the 1970s that the importance of fire on the landscape became recognized again. At that time, conservation and land management efforts increased, primarily focusing on the use of prescribed burns. Yet, this came with many challenges. First, fire had been excluded from this region for nearly 300 years, fundamentally changing the landscape they wished to burn. Furthermore, wildfires started to happen too frequently in the fynbos, damaging the habitat and putting nearby communities at risk. The region had also experienced significant landscape conversion and habitat fragmentation from urban and agricultural expansion (which also led to more human-related ignitions). Additionally, highly flammable invasive alien species like European pines and Australian wattles had taken over large swaths of the region. These issues, combined with a shortage of funding, safety concerns, and logistical issues related to multiple landowners has made fire management in the region challenging. From the 1970s to the 2010s, it is estimated that only 10% of the landscape had been actively managed.[9]

Currently, this region of South Africa is promoting an integrated and adaptive fire management approach focused on conserving the sensitive fynbos ecosystem while working to protect nearby communities from wildfire risk. To do this, they are relying on suppressing high-intensity dry-season fires while safely burning low-intensity wet-season fires.[10]

Figure 2.2 (*Left*) The MTC region of the Cape showing the two Mediterranean zones as defined by the Köppen climate classification: Csb and Csa.

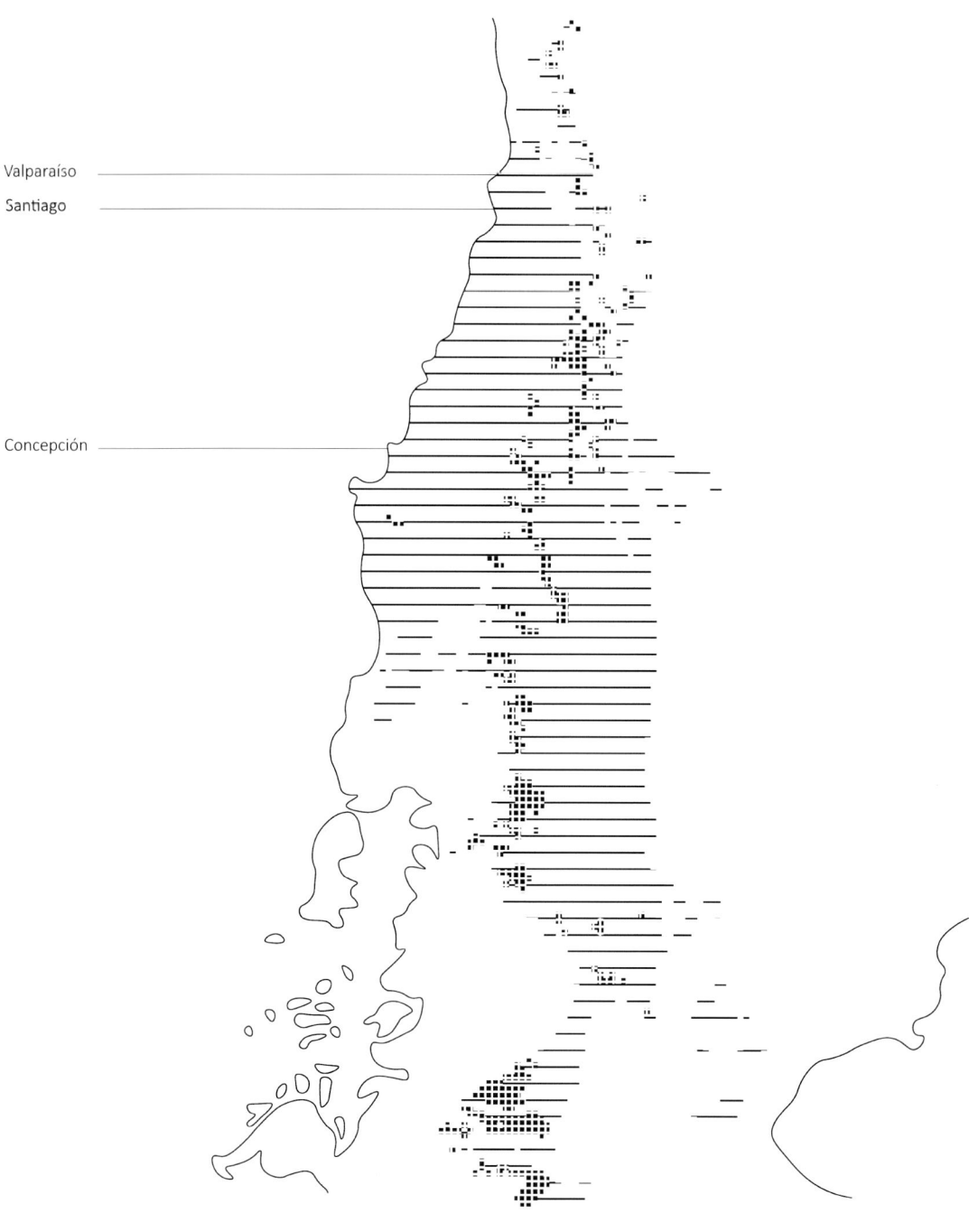

Valparaíso

Santiago

Concepción

Csb- Warm-summer Mediterranean climate

Csa- Hot-summer Mediterranean climate

100 mi

# B

## *Central Chile*

The MTC area of Central Chile, located on the western coast of South America, includes the cities of Santiago, Valparaíso, and Concepción. Topographically, it is defined by a coastal mountain range to the west and the Andes mountains to the east with a large central valley in between.

A majority of the central valley is human-modified for agricultural purposes – primarily for fruit and wine production – with some savanna. The dominant vegetation type at lower elevations is an evergreen shrubland called *matorral* and at higher elevations, the shrubland transitions into woodland communities. According to the National Forest Corporation of Chile, about 4% of Chile's total land area is devoted to non-native tree plantations comprising primarily eucalyptus and pine for wood production[11]; this is even more pronounced in Central Chile where it is estimated that 20% of the region is used for plantations.[12] These plantations often displace native woodland habitat, which is a priority ecosystem for conservation.[13]

Up until the Andean uplift, Chile was naturally fire-prone with summer lightning storms moving into the region to ignite the landscape. After this tectonic shift, the number of naturally occurring wildfires decreased dramatically.[14] Then, following the rise of colonization in the mid-1500s, there was a significant increase in human-induced wildfire activity due to fire being used as a tool to prepare the landscape for agricultural use, the pushing of development into wildlands making human-ignited events more common, and the planting of non-native species across the landscape. More recently, since 1985, the number of wildfires across the country has grown even more, especially in Central Chile with its unique climate and development patterns.[15] In this region, there are a large number of forestry plantations with highly flammable species, high incidences of arson, significant growth in the wildland-urban interface (WUI), high fuel load accumulation, and suppression complications due to challenging topography. As a result, wildfires have posed increased threats to a range of communities, led to a loss of native habitat, and negatively impacted forestry operations.

Today, there is a push toward wildfire risk assessments and large-scale planning initiatives in Central Chile. Instead of focusing solely on suppression efforts, there is an acknowledgement that wildfires will happen and that resources should be reserved for particularly high-intensity fires with destructive potential and for areas with significant risk. As part of this work, there is a push to diversify the forestry sector, reduce agricultural burning, and manage fuel in the WUI through pruning, clearing, and the construction of firebreaks.[16]

Figure 2.3 (*Left*) The MTC region of Central Chile showing the two Mediterranean zones as defined by the Köppen climate classification: Csb and Csa.

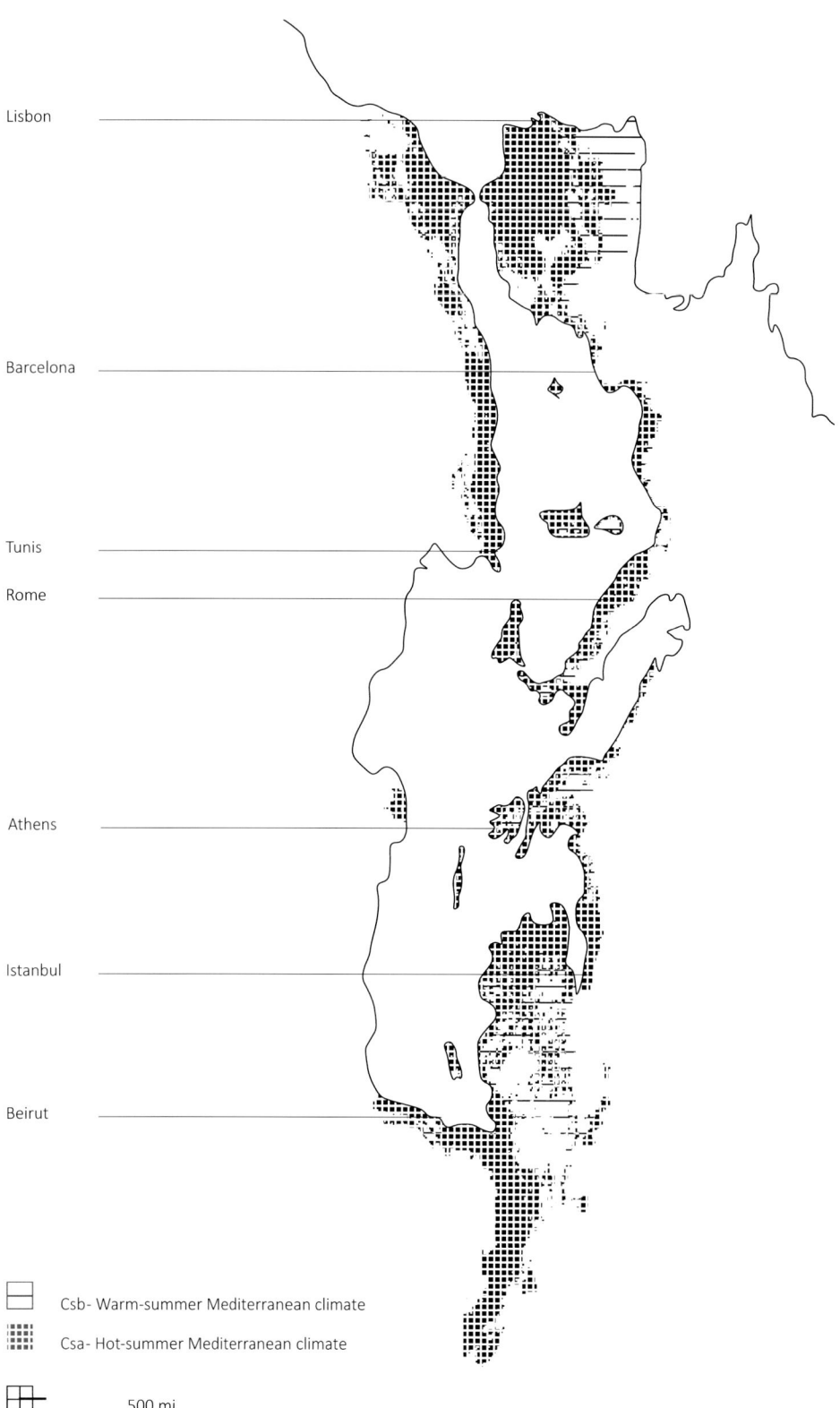

Lisbon

Barcelona

Tunis

Rome

Athens

Istanbul

Beirut

Csb- Warm-summer Mediterranean climate

Csa- Hot-summer Mediterranean climate

500 mi

# C

## *The Mediterranean Basin*

The MTC zone of the Mediterranean Basin, located on the periphery of the Mediterranean Sea between two major land masses, is a large and diverse geography spread across 20 coastal countries. It contains the following cities: Lisbon, Barcelona, Rome, Athens, Istanbul, Beirut, Tel Aviv, Algiers, and Tunis. Topographically, this large region is defined by mountainous terrain and a varied shoreline.

Historically, the Mediterranean Basin was dominated by oak, deciduous, and coniferous forests, but these have largely disappeared due to extensive human modification of the landscape. Today, the basin is considered a biodiversity hotspot and hosts a high number of endemic species due to its varied topography and management. There are three main vegetation types in the region. First, there is the broad-leaved evergreen shrubland, which is dominant. This type of vegetation, also called *maquis, macchia, gariga, phrygana, tommillares*, and *batha*, requires management either through grazers or periodic fires. Then, there is the *garrigue* which typically has soft leaved and drought-resistant plants. Lastly, there is the forested vegetation type which, while limited, can still be found in the region. Over time, these habitats have undergone significant changes due to shifts in land use, development, and fire regimes.[17]

Fires have always been a natural and ecologically important part of the Mediterranean Basin. Ever since the initial occupation of the region, humans have used fire to manipulate the landscape or allowed lightning-ignited fires to burn in an effort to support agropastoral activities. Yet, over the last several decades, a rural exodus in the Basin has led to an abandonment of the agricultural edge, catalyzing an expansion of highly flammable early successional species. This has, in turn, increased fuel loads in the fire-prone region, contributing to larger, more frequent, and more destructive fires. Up until recently, wildfire management in the basin has largely focused on short-term and reactive suppression strategies.[18]

Today, there is a collective understanding that fires are an endemic and necessary part of the Mediterranean Basin landscape and that there should be a focus on integrated fire management over suppression. Regionally, there are longer-term efforts underway to create more fire-resilient landscapes through preventative actions. These include diversifying the forests, creating bioeconomies through fuel reduction activities, and applying prescribed burns. There is also a push to protect communities that are at a high risk of loss from wildfires.

Figure 2.4 (*Left*) The MTC region of the Mediterranean Basin showing the two Mediterranean zones as defined by the Köppen climate classification: Csb and Csa.

Perth

Adelaide

Csb- Warm-summer Mediterranean climate

Csa- Hot-summer Mediterranean climate

500 mi

# D

## *South and Southwest Australia*

The MTC regions of South and Southwest Australia sit along the Indian Ocean coast and are home to the cities of Adelaide and Perth. Topographically, these areas are defined by a few low mountain ranges as well as a series of plateaus and valleys, with old, weathered, and low-nutrient soils.

Due to its topographic diversity as well as its soil, the region has highly diverse vegetation and has a number of endemic species. The three primary plant communities in the region are the *kwongan* which is a heathland, the *mallee* which is a shrubby woodland, and an evergreen forest.[19] Following European settlement in the early 1800s, though, these plant communities were significantly modified and fragmented due to agricultural production (primarily wheat), large-scale grazing efforts, urban expansion, deforestation, and the introduction of invasive alien species. It is estimated that over 50 plant species have already gone extinct with over 200 currently threatened.[20]

Prior to European settlement of this region, it was common for the landscape of South and Southwest Australia to burn. Lightning fires naturally lit the landscape and were allowed to burn out on their own and indigenous groups intentionally lit low-intensity fires to manage the landscape. In the wake of colonization, though, fire suppression became the norm to protect communities as well as areas of agricultural and forest-related production. In these areas, extensive firebreaks were constructed into a series of blocks to support suppression activities. Despite these efforts, the frequency of high-intensity wildfires in this region increased, and forest health declined. In time, officials began understanding the importance of rotational burning to manage the landscape, so in the 1960s, they introduced policies focused on hazard reduction through prescribed burning. While this burning succeeded in reducing some risk, the amount of fuel in the landscape made these efforts dangerous. Furthermore, many conservation groups challenged the burns claiming that they were threatening local biodiversity.[21]

Today, in the regions of South and Southwestern Australia, there is still a large focus on fire suppression to reduce the risk of catastrophic events. And while hazard reduction burns are still common, they are not practiced widely enough from year to year. In recent years, there has been a push toward diversifying wildfire management to include controlled burns focused on ecological regeneration and cultural burns implemented by indigenous fire groups. With both of these efforts, the goal is to create intentionally patchy burns with variations in season, frequency and intervals, to promote habitat heterogeneity. These regions are also exploring other wildfire management options including pruning, grazing, and promoting native vegetation that is fire-resistant.

Figure 2.5 (*Left*) The MTC region of South and Southwestern Australia showing the two Mediterranean zones as defined by the Köppen climate classification: Csb and Csa.

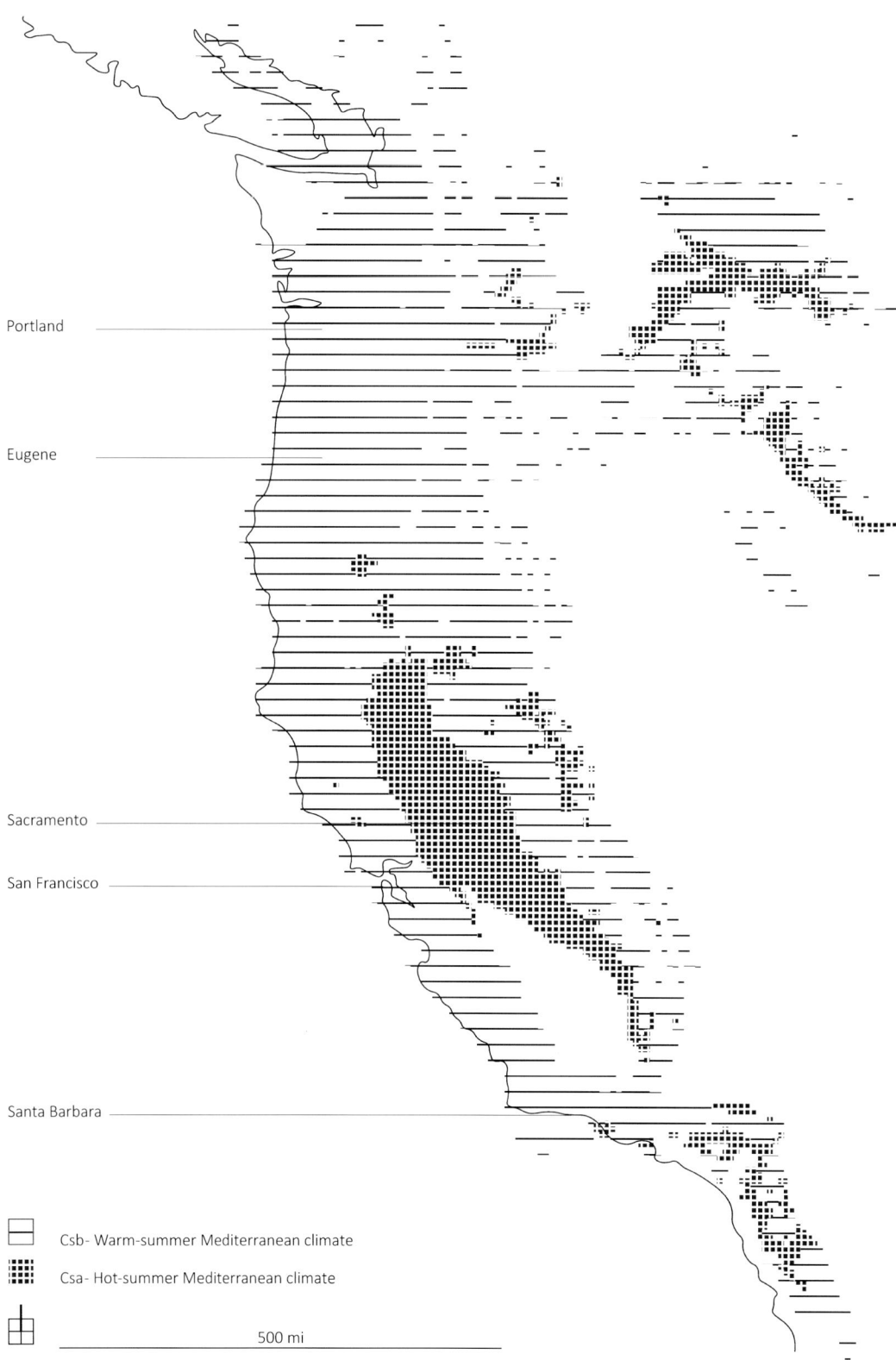

Portland

Eugene

Sacramento

San Francisco

Santa Barbara

Csb- Warm-summer Mediterranean climate

Csa- Hot-summer Mediterranean climate

500 mi

E

## Western North America

The MTC region of Western North America is situated near the Pacific Ocean and includes the cities of Portland, Eugene, Sacramento, San Jose, Santa Barbara, and Los Angeles. Topographically, it is complex and young, and includes coastal ranges, wide valleys, and foothills.[22]

Historically, there were four main vegetation types in this region: grasslands with perennial bunchgrasses and annuals, savannas dominated by oaks, chaparral with broad-leaved evergreen shrubs, and coastal sage scrub with soft, drought-deciduous plants. With the arrival of Europeans in the 1800s, these habitats began to shift and change. Urbanization, agricultural cultivation, and grazing began fragmenting these landscapes and introducing a range of invasive alien species. Some vegetation types, like grasslands, now cover merely 10% of their historical range due to significant land use changes over the past 200 years.[23]

Prior to colonization, indigenous populations used low-intensity fire as a tool to shape the landscape and actively allowed lightning-ignited fires to burn. Yet, beginning in the late 1800s, fire suppression became a widely adopted practice in the region resulting in fuel buildups in forested areas as well as a range of landscape conversions. This strict policy of fire exclusion persisted until the 1960s when wildfires became recognized as an important ecological (and cultural) process. Over the last few decades, there has also been an expansion of the WUI in Western North America, a trend which has not only increased the risk of ignition but has put more people in wildfire-prone areas.[24]

Today, Western North America has experienced a number of destructive wildfires and has allocated a significant amount of public funding toward suppression efforts to protect existing communities and sensitive infrastructure. At the same time, there has also been a push toward more preventative and landscape management-based initiatives like fuel reduction activities, planned burns – both controlled and cultural – and the allowance of some unplanned wildfires to freely burn across landscapes.

In the following chapters, we explore a wide range of land-fire stewardship practices from these different regions, revealing both similarities and differences in design.

Figure 2.6 (*Left*)
The MTC region of Western North America showing the two Mediterranean zones as defined by the Köppen climate classification: Csb and Csa.

# Notes

1    John Keeley, "Fire in Mediterranean Climate Ecosystems – A Comparative Overview," *Israel Journal of Ecology and Evolution* 58, no. 2 (May 2013): 123–135.
2    Philip Rundel et al., "Fire and Plant Diversification in Mediterranean-Climate Regions," *Frontiers in Plant Science* 9, no.851 (July 2018): 1–13.
3    Gloria Montenegro et al., "Fire Regimes and Vegetation Responses in Two Mediterranean-Climate Regions," *Revista Chilena de Historia Natural* 77 (2004): 455–464.
4    Alexandra Syphard, Volker Radeloff, Todd Hawbaker, and Susan Stewart, "Conservation Threats Due to Human-Caused Increases in Fire Frequency in Mediterranean-Climate Ecosystems," *Conservation Biology* 23, no.3 (2009): 758–769.
5    Ioannis N. Vogiatzakis, Antoinette Mannion, and Guy Griffiths, "Mediterranean Ecosystems: Problems and Tools for Conservation," *Progress in Physical Geography* 30, no.2 (2006): 175–200.
6    Montenegro et al., "Fire Regimes."
7    Syphard, Radeloff, Hawbaker and Stewart, "Conservation Threats."
8    Brian van Wilgen, "Fire Management in Species-Rich Cape Fynbos Shrublands," *Frontiers in Ecology* 11, no.1 (2013): e35–e44.
9    Ibid.
10   Ibid.
11   Xavier Ubeda, and Pablo Sarricolea, "Wildfires in Chile: A Review," *Global and Planetary Change* 146 (October 2016): 152–161.
12   Vogiatzakis, Mannion, and Griffiths, "Mediterranean Ecosystems."
13   Ubeda and Sarricolea, "Wildfires in Chile."
14   Keeley, "Fire in Mediterranean."
15   Ubeda and Sarricolea, "Wildfires in Chile."
16   Ubeda and Sarricolea, "Wildfires in Chile."
17   Juli Pausas, Joan Llovet, Anselm Rodrigo, and Ramon Vallejo, "Are Wildfires a Disaster in the Mediterranean Basin? – A Review," *International Journal of Wildland Fire* 17 (2008): 713–723.
18   Ibid.
19   Richard Cowling et al., "Plant Diversity in Mediterranean-Climate Regions," *TREE* 11, no.9 (September 1996): 326–366.
20   Vogiatzakis, Mannion, and Griffiths, "Mediterranean Ecosystems."
21   Vic Jurkskis, Bob Bridges, and Pual de Mar, "Fire Management in Australia: The Lessons of 200 Years." In *Joint Australia and New Zealand Institute of Forestry Conference Proceedings*, Queenstown, 2003, 353–368. Wellington: Ministry of Agriculture and Forestry.
22   Cowling et al., "Plant Diversity."
23   Vogiatzakis, Mannion and Griffiths, "Mediterranean Ecosystems."
24   Syphard, Radeloff, Hawbaker and Stewart, "Conservation Threats."

# Bibliography

Cowling, Richard, Philip, Rundel, Byron Lamont, Mary Arroyo, and Margarita Arianoutsou, "Plant Diversity in Mediterranean-Climate Regions." *TREE* 11, no.9 (September 1996): 326–366.

Jurkskis, Vic, Bob Bridges, and Pual de Mar, "Fire Management in Australia: The Lessons of 200 Years." In *Joint Australia and New Zealand Institute of Forestry Conference Proceedings*, Queenstown, 2003, 353–368. Wellington: Ministry of Agriculture and Forestry.

Keeley, Jon, "Fire in Mediterranean Climate Ecosystems – A Comparative Overview." *Israel Journal of Ecology and Evolution* 58, no.2 (May 2013): 123–125.

Montenegro, Gloria, Rosanna Ginocchio, Alejandro Segura, Jon Keely, and Miguel Gomez, "Fire Regimes and Vegetation Responses in Two Mediterranean-Climate Regions." *Revista Chilena de Historia Natural* 77 (2004): 455–464.

Pausas, Juli, Joan Llovet, Anselm Rodrigo, and Ramon Vallejo, "Are Wildfires a Disaster in the Mediterranean Basin? – A Review." *International Journal of Wildland Fire* 17 (2008): 713–723.

Rundel, Philip, Mary Arroyo, Richard Cowling, Jon Keeley, Byron Lamont, Juli Pausas, and Pablo Vargas, "Fire and Plant Diversification in Mediterranean-Climate Regions." *Frontiers in Plant Science* 9, no.851 (July 2018), 1–13.

Syphard, Alexandra, Volker Radeloff, Todd Hawbaker, and Susan Stewart, "Conservation Threats Due to Human-Caused Increases in Fire Frequency in Mediterranean-Climate Ecosystems." *Conservation Biology* 23, no.3 (2009), 758–769.

Ubeda, Xavier and Pablo Sarricolea, "Wildfires in Chile: A Review." *Global and Planetary Change* 146 (October 2016), 152–161.

Vogiatzakis, Ioannis, Antoinette Mannion, and Guy Griffiths, "Mediterranean Ecosystems: Problems and Tools for Conservation." *Progress in Physical Geography* 30, no.2 (2006), 175–200.

Wilgen, Brian van, "Fire Management in Species-Rich Cape Fynbos Shrublands." *Frontiers in Ecology* 11, no.1 (2013), e35–e44.

# Pyric Lexicon

Figure 3.1

View of a landscape around Lake Berryessa after the LNU Lightning Complex Fires of 2020.

# Chapter 3

# Pyric Lexicon

*The following terms are underlined across the book as key fire-related concepts:*

| | |
|---|---|
| Active sheltering | systematically monitoring interior and exterior conditions during a wildfire event and actively working to prevent loss of property and life |
| Aerial ignition | dropping fire-producing devices from aircrafts to ignite prescribed fire |
| Anthropogenic fire | intentional and unintentional fire created by humans |
| Backburning | strategically igniting an area ahead of an approaching fire to reduce the amount of vegetation |
| Backing fire | a fire moving against the wind or downslope that typically burns slowly and at a low-intensity |
| Block burning | the rotational planned burning of subdivided landscape areas |
| Burn boss | an individual who is certified to lead planned burns |
| Burn plan | a written document outlining the details for a planned burn including weather restrictions, personnel, and equipment |
| Burn window | the time when environmental conditions are appropriate or legally permitted for a planned burn |
| Bushfire | an unplanned vegetation fire typically located in Australia |
| Combustible | capable of igniting and burning |
| Community refuge | an area designated as a safe space if fire evacuation is not possible |

## Introduction

Community wildfire protection plan
: a document that identifies risks and potential actions that a community could take in the face of wildfire

Controlled burn
: a planned vegetation fire

Conductive heat
: when heat moves from one entity to another through direct contact

Convective heat
: when heat moves from one entity to another through rising hot air

Cool burn
: a planned slow-moving and low-intensity surface fire, typically occurring outside of peak fire season

Crown fire
: a fire that has reached into a forest canopy and spreads from tree to tree

Cultural burn
: an intentional fire set by indigenous people to manage the landscape and support culturally-important species

Defensible space
: the space around a building that is actively managed to reduce the risk of loss during a wildfire

Direct attack
: a firefighting technique that actively treats burning material

Edible fire buffer
: an area planted with low-flammability crops to protect properties from wildfire

Ember
: a small piece of burning debris (also known as a "firebrand")

Ember attack
: when wind carries small pieces of burning debris and increases the potential for ignition

Ephemeral refugia
: an area of the landscape that remains unburned or lightly burned after one wildfire event

Fire-adapted community
: a community that works to identify risks and potential strategies for reducing the negative impacts of wildfire

Fire behavior
: the way in which material ignites, as well as how the fire develops and spreads

| | |
|---|---|
| Firebreak | a strip of land, typically with limited vegetation, managed to change the behavior of a fire or serve as a space for firefighting |
| Fire bunker | a structure designed to resist ignition where people can retreat to during a wildfire event |
| Fire corridor | an area of the landscape that is susceptible to repeat wildfire events |
| Fire curtain | an architectural barrier that prevents the spread of fire from one space to another |
| Fire danger index | a scale used for predicting fire ignition potential |
| Fire deficit | a shortage of expected or typical fire activity |
| Fire effects monitor | an individual responsible for recording information about a fire and communicating this information |
| Fire exclusion | suppressing all wildfires in an area |
| Fire front | an area of a fire that is actively burning or smoldering |
| Fire hazard severity zone | areas in California with significant fire hazards that have mandates for reducing risk |
| Fireline | a strip of land consisting of mineral soil that is constructed to change the behavior of an approaching fire or serve as a space for firefighting |
| Fire perimeter | the entire edge of a fire |
| Fire regime | the general pattern of how and when wildfires occur in an ecosystem |
| Fire-resistant | a material with inherent properties that make it difficult to ignite |
| Fire safe council | a community-based organization that works to reduce wildfire risks |
| Fire scar | a tree defect caused from excessive heating (also known as a "catface") |
| Fireshed | areas where fires typically ignite and are likely to spread to communities |

**Introduction**

| | |
|---|---|
| Fire shelter | a deployable safety device for entrapped firefighters |
| Fire spread | the way in which a fire moves through an area, typically in the ground, on the surface or through the canopy |
| Firestick farming | a planned fire set by indigenous people in Australia to manage the landscape and support culturally-important species |
| Fire severity | measures the impact of a fire on the landscape ranging from low- to high- to mixed-severity |
| Fire triangle | the three primary elements that are needed for ignition: an oxidizing agent, heat, and fuel |
| Fire weather | when environmental conditions are conducive to ignition including: low relative humidity, strong winds, and the potential for dry lightning |
| Firewise | a recognition program that helps communities reduce their wildfire risk |
| Firing boss | an individual responsible for leading ignition treatments |
| Flanking fire | a fire that burns parallel to the wind direction, set typically to connect a backing fire to a head fire |
| Fuel | any material that can burn in a fire including vegetation and structures |
| Fuelbreak | a strip of land with reduced vegetation to change the behavior of an approaching fire or serve as a space for firefighting |
| Fuel load | the amount of burnable material within a given unit |
| Fuel reduction | the removal or reduction of vegetation and combustible material |
| Good fire | planned fires intended to improve forest health and reduce wildfire risks |
| Hand crews | firefighting personnel who work on the ground to contain and suppress fires |
| Hand firing | igniting planned fires by hand |

| | |
|---|---|
| Handline | when vegetation is removed by hand from a strip of land |
| Hardening | retrofitting the vulnerable parts of structures to reduce the risk of damage |
| Head fire | a fire that burns with the wind |
| High-severity | when a fire burns so intensely that it kills most (over 80%) of the existing trees in a given area |
| Home ignition zone | the area 100–200 feet away from a structure that is particularly vulnerable during a wildfire |
| Hostile fire | a planned fire that becomes uncontrollable or expands beyond its boundaries |
| Hotshot crew | firefighting personnel that work on the hottest parts of wildfires |
| Hotspot | isolated pockets of the landscape that are actively burning or smoldering |
| Ignition-resistant | materials that resist ignition and sustained burning when exposed to heat or flames |
| Incident action plan | written or verbal plan for managing a wildfire event |
| Indigenous fire | an intentional fire set by native people to manage the landscape and support culturally-important species |
| Indirect attack | a firefighting technique that treats land ahead of a fire front |
| Invasive alien species | species that are outside of their natural range and threaten biological diversity; post-wildfire, these species are typically well-adapted for dispersal and growth |
| Ladder fuels | burnable material that can transfer a surface fire into a canopy fire |
| Live fuel moisture content | the amount of moisture found within plants |
| Low-severity | when a fire burns at a low intensity, killing few (under 30%) of the trees in a given area |

## Introduction

Management blocks          designated zones for rotational fire management

Mechanical thinning          using machinery to remove trees in an overgrown forest

Mixed-severity          when a fire burns a landscape in a variable pattern with patches of different intensities

Moderate-severity          when a fire burns and kills some (between 30 and 80%) trees in a given area

Mopping up          firefighting measures to minimize the spread of fire into unburned areas

Noncombustible          materials not known to ignite or burn; typically requiring a standardized test

Non-commercial thinning          removing trees in an overgrown forest that have little-to-no commercial value

Patch-mosaic burning          a planned fire strategy to create patchiness and heterogeneity across space and time

Persistent refugia          an area of the landscape that remains unburned or lightly burned after multiple wildfire events

Personal protection equipment          gear issued to firefighters to help reduce the potential for burns and other injuries

Piling and burning          when debris from thinning is distributed in piles in the forest and burned

Prescribed burn          an intentionally-set fire used for landscape management

Pre-suppression program          the training of firefighting personnel ahead of a wildfire event to promote safe, effective, and efficient strategies

Preventative forestry          the removal or reduction of vegetation in a forest to reduce wildfire risks

Priority protection perimeters          large areas of forested land in Spain that are at a high risk of wildfire

Pyrocene          a term coined by Stephen Pyne that describes a new epoch defined by the anthropogenic use of fire and an increase of fire events

| | |
|---|---|
| Pyrodiversity | how environmental conditions can affect the ways in which fire touches the landscape across space and time |
| Pyrogeography | a discipline focused on the past, present, and future spread of wildfire |
| Radial thinning | removing smaller trees that are growing under the canopy of larger trees |
| Radiant heat | when heat moves from one entity to another through air in all directions |
| Refugia | areas of the landscape that are unburned or minimally burned after a wildfire |
| Retardant | a substance used to slow down, stop, or reduce the intensity of wildfires by reducing the flammability of materials; it is typically comprised of water, fertilizer, and a colorant |
| Shaded fuelbreak | a strip of land with reduced surface vegetation to change the behavior of an approaching fire or serve as a space for firefighting; typically, these areas maintain some tree canopy |
| Shelter-in-place | staying in a structure, instead of evacuating, during a wildfire |
| Size-up | the continuous evaluation of fire conditions during firefighting efforts |
| Sling loads | when resources are delivered aerially to firefighters in remote areas |
| Smoke jumper | specially trained firefighters who descend into remote locations via parachute |
| Spotting | when small pieces of burning debris are carried downwind of the main fire and start new fires |
| Staging area | a safe command post for firefighters to report and receive supplies |
| Strategic management points | areas in the landscape that help to slow, stop, or reduce the intensity of wildfires |

## Introduction

| | |
|---|---|
| Surface fire | a fire that burns on the ground |
| Temporary refuge area | a predetermined area where people can safely retreat to during a wildfire event |
| Thinning | removing vegetation in an overgrown forest |
| Thinning-from-below | removing small trees and lower branches from the ground in an overgrown forest |
| Voluntary buyout program | residents affected by wildfire receive money for their lost property which is then turned into open space |
| Vulnerability assessment | a tool used to assess potential wildfire risks and vulnerabilities |
| Wildfire detection systems | a system used to detect and report wildfires to local authorities |
| Wildfire risk reduction buffers | a strip of land with reduced vegetation to change the behavior of an approaching fire, serve as a space for firefighting, or function as open space |
| Wildfire suppression | the act of extinguishing vegetation fires |
| Wildland-urban interface (WUI) | the space between wildland and the built environment |
| Wind-driven fire | when wind accelerates the spread of a fire |

# Approaches to Designing with Fire

# Resistance

Figure 4.1
View of a landscape and trail around Lake Berryessa (Northern California) after the LNU Lightning Complex Fires of 2020.

Figure 4.2
View of a highly dense conifer forest. Photograph by Derek Young.

Figure 4.3
A driveway is all that remains of a house in Berryessa Highlands, a ridgeline community devastated by the LNU Lightning Complex fires of 2020. In the background, a new house is being rebuilt.

Figure 4.4
Effects of the 2012 King Fire on an untreated forest. Photograph by Malcolm North.

Figure 4.5
The framing of a new house goes up in Berryessa Highlands, a ridgeline community devastated by the LNU Lightning Complex fires of 2020.

Figure 4.6
Firefighters mopping up by mixing water with dirt to extinguish any remaining flames. Photograph by Washington Southeast DNR.

# Chapter 4

# Resistance

Resistance

*to exert force in opposition; to withstand the force or effect of; to fight, combat, oppose; to strive against.*

These are approaches that tend to take a stand against the creative and transformative propensities of fire and forces of landscape change. To resist in this context, is to try to control, or to maintain the overall status quo of landscapes and how they are used and occupied. In contemporary applications, resistance is often characterized by the continuation and augmentation of established design, policy and management techniques that are increasingly challenged by the effects of past efforts (whether intended or not), as well as new and emergent challenges brought about by a rapidly changing climate and pervasive landscape alteration.

DOI: 10.4324/9781003172956-6

1 mi

Yosemite Valley

El Capitan

01

*Firefighting*

Half Dome

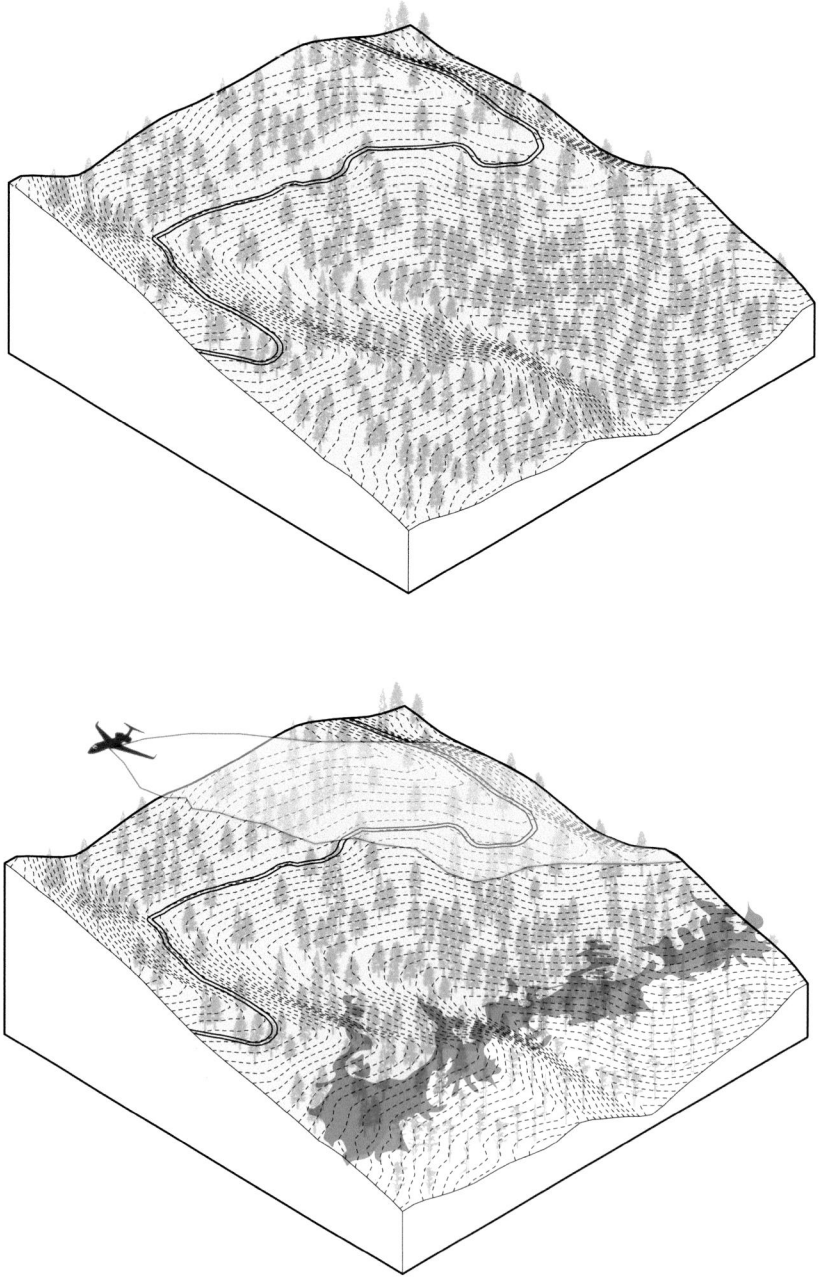

01

*Firefighting*

| | |
|---|---|
| Description | Preventing the occurrence of natural fire and <u>anthropogenic fire</u> on the landscape |
| Example Location | Yosemite National Park, CA (and in California's Sierra Nevada montane forests more broadly) |
| Size | 747,956 acres |
| Primary Implementer | National Park Service |
| Team | American Red Cross, U.S. Forest Service, California Highway Patrol, and National Weather Service |

## Technique Overview

According to many studies, fires were common in Yosemite and throughout California's Sierra Nevada forests prior to Euro-American settlement. During this time, lightning-ignited fires burned across the region and indigenous residents of the area intentionally set fires to promote the growth of certain plants and to maintain a mosaic-like landscape. In fact, researchers studying mud cores in the Valley found an increase in ash after indigenous occupation.[1] These indigenous fires that burned prior to the mid-1800s were primarily low intensity and burned close to the ground.[2]

Then, when Euro-American settlers arrived, the number of fires in Yosemite dramatically decreased due to a colonial belief that the fires were damaging the landscape. In 1891, the practice of putting out fires became solidified when the U.S. Army was given responsibility for protecting the park. Twenty years later, following an unprecedented 1910 fire season, the federal government enacted a national policy promoting <u>fire exclusion</u> and, in time, Yosemite National Park became a model for suppression. With funding and resources from the New Deal, they built lookouts and other firefighting infrastructure, and trained slews of civilians to suppress conflagrations. By 1935, when the Forest Service enacted their 10am rule – that all wildfires needed to be extinguished by the morning after detection – authorities at Yosemite National Park were primed and ready to defend the landscape.[3]

For decades, authorities worked to control any active flames in the park by containing <u>fire perimeters</u>, and carefully extinguishing any burning material. Their goal was to stop small conflagrations before they had time to grow in size and intensity, and their plan worked.[4] By the mid-1900s, authorities suppressed fires so well that the annual area burned was a mere fraction of what it was historically.[5]

Then in the 1960s, authorities began realizing the negative effects of suppression in the park. At this time, Starker Leopold led a commission which advised park authorities to reintroduce fire to the landscape and actively manage the forests and they agreed. Then, in the late 1970s, the Forest Service followed suit and abandoned their <u>fire exclusion</u> policy.[6]

Despite this positive shift in policy, the damage associated with decades of fire suppression had already transformed the landscape. Without fire on the land, the forests in Yosemite became denser and young trees began encroaching in on meadows. This biomass then became <u>fuel</u> for larger, more destructive wildfires in the park.[7]

Figure 4.7 (*Previous*) Aerial of Yosemite Valley. Source: Google Earth 2022.

Figure 4.8 (*Left*) A view of Yosemite prior to a wildfire (top) and the same view with firefighting techniques being implemented to prevent the occurrence of any fire on the landscape (bottom).

Today, the park currently has two zones. One zone comprises 83% of the park and is an area where lightning-ignited fires are allowed to burn under the right conditions. The other zone comprises 17% of the park and is an area where all human- and lightning-ignited fires are suppressed due to development or proximity to the park's boundaries.[8]

## Planning and Design Process

Firefighting typically involves the following steps.

Following the detection of a new wildfire, the first step of fire suppression involves a process called size-up. This generally happens prior to arriving at an incident but also happens continuously through a wildfire event. During a size-up, all firefighting personnel gather information about the incident and monitor existing conditions to understand possible future conditions, including threats to lives and property.[9]

After a size-up, a leading authority develops an incident action plan (IAP) that focuses on saving lives, containing the fire, and saving property. The IAP has a list of operations for each of the three goals and should help authorities determine where they are succeeding and where they need to adjust resources. Some operations involve direct attacks – when firefighters act on the fire front – and some operations involve indirect attacks – when firefighters act away from the main fire front.[10]

There are seven main suppression operations typically employed in Yosemite National Park. The first technique involves firefighting personnel using hand tools to dig down to mineral soil to build firebreaks and stop an advancing fire. These handcrews also patrol the perimeter of the fire and cut down burning trees. In general, handcrews are helpful in areas without great vehicular or water access. The second technique involves heavy machinery to move earth and create firebreaks. This machinery can also be used to create access routes and to remove trees or shrubs ahead of a fire front. While this technique can expedite firefighting efforts, vehicular access is necessary. If there is access to surface water like a lake or pond, hoseline crews can be employed. These crews use pumps and hoses to siphon water from existing water sources to douse flames. The crews can also be used to wet vegetation ahead of a fire, protect property, and aid with backburning. If there is little to no access to surface water, ground tankers can be used; firefighting personnel can attach hoselines to trucks carrying water to help put out flames. Like the hoseline crews, this technique can also be used for wetting vegetation, protecting property, and backburning. If vehicular and ground access is limited, fire suppression can also happen aerially. Helicopters can drop suppressants and retardants on the fire front through either sling loads or bellytanks. Fixed wing planes called air tankers can also do this along with helping to transport crew members and firefighting equipment. Lastly, aircrafts can be used for aerial ignition, detection, and directing fire suppression efforts.[11]

## Performance and Evaluation

Under ideal conditions, fire suppression can protect lives and vulnerable buildings and infrastructure in fire-prone regions like Yosemite National Park. And it is a technique that will inevitably need to be used in some form in the future. It is obviously appealing for any community in the direct path of an aggressive <u>fire front</u>. Furthermore, it is appealing for risk-adverse communities where the social acceptance of <u>prescribed burns</u> is low and for communities unwilling to modify their built environment or surrounding landscape.[12]

Fire suppression is also a technique that immediately gratifies residents (unlike other techniques that may take months or even years to implement). Additionally, media outlets often frame firefighting as a war against flames with firefighting personnel as heroes. This portrayal reinforces fire suppression as a positive act, leading to increased support from the general public.

## Challenges and Future Research

Unfortunately, fire suppression is not always successful when attempting to protect lives, property, and infrastructure. This is partially because there is so much variability in wildfire events – from weather to <u>fuel</u> to terrain, every fire is unique. Furthermore, every incident is handled differently depending on who is directing the IAP. Lastly, while small fires are easier to suppress, many wildfires grow rapidly in size and become very difficult to manage in a relatively short period of time.[13]

Fire suppression can negatively affect overall forest health. In the Sierra Nevada, many forests need low-intensity fire to stay healthy and to limit the number of trees and plants competing for scarce resources like water. Without regular fires, these forests can become so overgrown that a high-intensity wildfire can take out most of the vegetation.[14]

Fire suppression is extremely expensive and resource-intensive due to the number of firefighting personnel and equipment needed to battle wildfires. There is also the potential issue of labor shortages due to low pay and dangerous work conditions. And while many safety measures are in place to protect firefighters, they may be pressured to take on more risk in extreme situations.

Additionally, in order for fire suppression to be successful, there needs to be a strong <u>pre-suppression program</u> for educating the general public, training firefighting personnel, obtaining and maintaining suppression equipment, and honing <u>wildfire detection systems</u>.[15]

Lastly, fire suppression is a catch-22 – due to decades of historical fire suppression in the Sierra Nevada, the forests now contain too much <u>fuel</u>. While this fuel needs to burn, doing so creates too much risk, so fire suppression is used and the amount of fuel in the forest just continues to grow. This buildup of fuel has the potential to lead to larger and more severe wildfires.[16] To halt this trend, many researchers believe there needs to be a shift in wildfire policy to better balance the use of reactive measures like fire suppression with proactive measures to mitigate fire risk ahead of an event.[17]

1 mi

Giant Forest

*Foil Wrapping*

Sequoia National Park

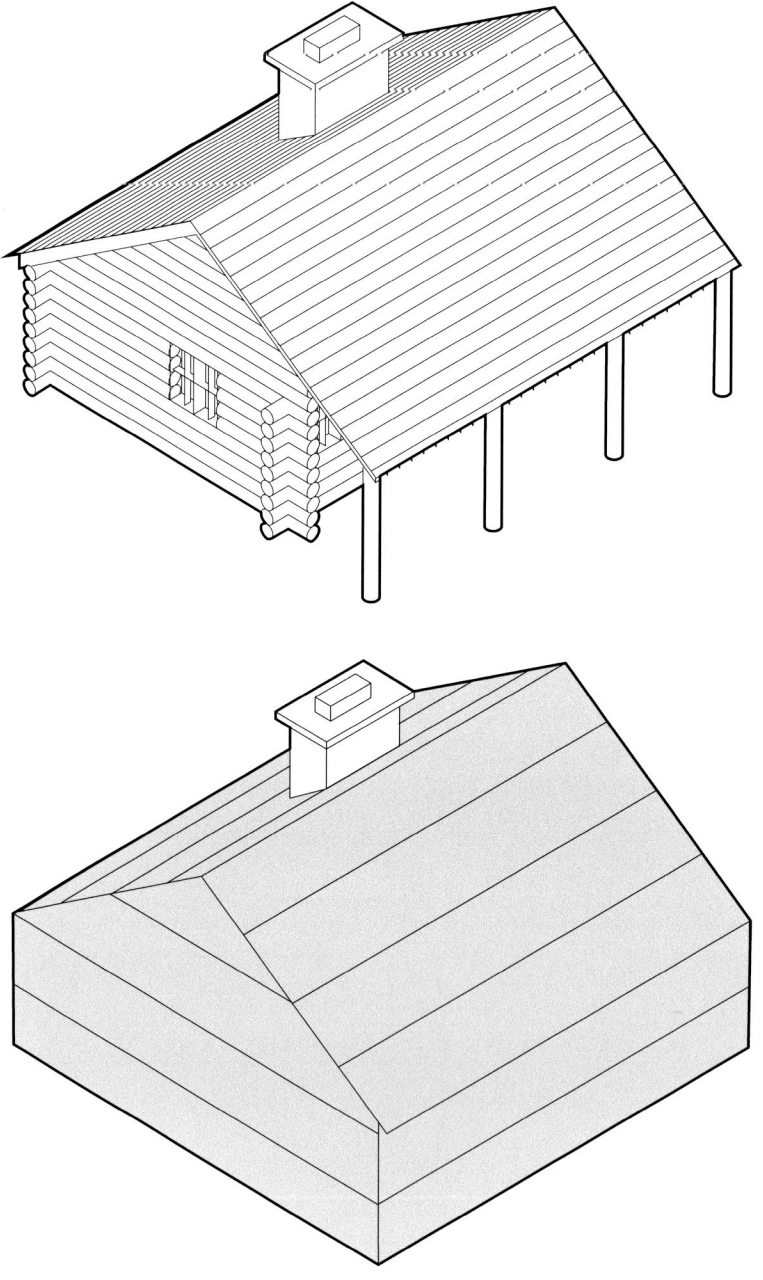

02

*Foil Wrapping*

| | |
|---|---|
| Description | Wrapping structures and heritage trees in aluminized blankets |
| Example Location | Sequoia National Park, CA, United States |
| Size | 1,880 acres |
| Primary Implementer | United States Forest Service |
| Team | National Park Service |

## Technique Overview

Since 1977, every federal wildland firefighter in the U.S. has been required to carry a <u>fire shelter</u> in the case of an emergency. If a wildfire entraps a firefighter and they cannot escape without potential injuries, they can deploy the shelter in a safe location, lie down, and wait for the <u>fire front</u> to pass. The personal fire shelter model has three primary layers: an outer layer made of aluminum and silica cloth that reflects <u>radiant heat</u> and slows heat transfer, an intermediary air gap layer that insulates, and an interior layer made of aluminum and fiberglass that stops heat from re-radiating.[18]

Firefighters used this technology primarily for <u>personal protection equipment</u> until 1988, when wildfires threatened Yellowstone National Park. At this time, firefighters decided that they could not safely protect historical structures so they deconstructed their fire shelters and stapled the material to the buildings. To their surprise, the foil wrapping technique worked and the buildings did not burn. Since then, a number of national agencies including the United States Forest Service (USFS) and the Bureau of Land Management (BLM) have used aluminized blankets to protect important structures and infrastructure in the line of fire.[19]

In recent years, the demand for this material has significantly increased and the market has expanded beyond federal agencies to include private entities such as resorts and homeowners in fire-prone areas like the <u>wildland-urban interface (WUI)</u>.[20] One reason for this expansion is the relatively new finding that firebrands and structure-to-structure spread are significant causes of ignition. In an attempt to protect individual structures from these vulnerabilities, many are turning to fire blankets.[21]

In September 2021, the USFS tested this technique on giant sequoias, some of the world's largest and oldest trees. At this time, the KNP Complex Fire was approaching California's Sequoia National Park and firefighters wanted to protect the trees any way they could. (While giant sequoias are fire-adapted and actually need fire to reproduce, high intensity events like the KNP Complex can severely damage or kill trees.)[22] Along with raking up leaves, rolling away logs, digging duff into the ground, dispersing fire <u>retardant,</u> and building <u>firelines</u>, the team wrapped the bases of the trees in aluminized blankets typically used for structural protection.[23]

## Planning and Design Process

Similar to the personal <u>fire shelters</u> used by firefighting professionals, aluminized blankets offer fire protection in three ways: they prevent firebrands from entering gaps and igniting,

Figure 4.9 (*Previous*) Aerial of the Giant Forest in Sequoia National Park. Source: Google Earth 2022.

Figure 4.10 (*Left*) A view of a historical structure prior to a wildfire (top) and the same view with aluminum blankets wrapped around the structure to reflect radiant heat and prevent ember intrusion (bottom).

they block direct contact with flames, and they reflect radiant heat.[24] They are typically made of Kevlar, aluminum laminate, and fiberglass and come in a roll or folded sheets as large as 5'x300'. Aluminized blankets can also be customized to fit the structure they are protecting. Installers can use staples or carbon steel bands with pre-drilled holes to attach the blankets; this should happen at least 30 minutes ahead of a fire front.[25]

To protect houses, fire lookouts, bridges, power line poles, railings, structural supports, fences, and equipment, aluminized blankets should ideally cover and enclose each structure fully. This will prevent firebrands from entering vulnerable gaps like eaves, vents, and gutters.[26] If installers do not have enough material to cover the entire structure, they should prioritize and wrap the most vulnerable parts of the structure. For instance, if a house has an untreated wood shake-and-shingle roof, this should be wrapped first.

To protect heritage trees like giant sequoias, aluminized blankets should cover the base of each tree and any fire scars. The base of the tree should be prioritized for two reasons. First, it is logistically challenging to wrap an entire tree in aluminized blankets, especially in an emergency. Second, the most vulnerable part of most trees (especially those with thick, fire-resistant bark or those that have been adequately pruned and maintained) is their canopy. Therefore, wrapping the base helps to prevent a surface fire from becoming a crown fire, which can easily spread from tree to tree. Covering existing fire scars should also be prioritized, as these can be vulnerable openings for firebrands to ignite.[27]

## Performance and Evaluation

While the evidence for the efficacy of aluminized blankets is mostly anecdotal, a few studies have shown that the material can block over 90% of convective heat and radiant heat.[28]

Aluminized blankets are also considered fast to deploy. Installers can cover a roof in just minutes with a large panel and an average-sized house in five to six hours. Furthermore, installation does not require professional training, unlike many conventional wildfire suppression techniques.[29] Installers can also deploy the blankets way ahead of a fire front and take them down long after a wildfire has passed. This also differs from foams, gels, and water sprays which need to be applied right before an event and cleaned up immediately after. By having a longer installation window, aluminized blankets put less of a burden on firefighters and homeowners since they can be installed before conditions get too dangerous.[30]

Aluminized blankets do not require any water or power, which can be helpful in remote locations or if emergency power shut-offs occur. Foams, gels, and water spray suppression systems often need both to function properly.

Lastly, while many conventional wildfire suppression techniques can only be used once, aluminized blankets are reusable, with a shelf life of 10–15 years.[31]

## Challenges and Future Research

Again, while there is significant anecdotal evidence and manufacturer disseminated test results showing that aluminized blankets protect structures and trees from wildfire, there is little documentation in scientific literature. In addition, there is a dearth of performance-based standards and third-party certification for the blankets.[32]

The success of aluminized blankets during wildfire events also heavily depends upon the precision by which they are installed. If installed hastily or improperly, the blankets can tear, seams can open, and heat can get trapped. These issues can make the structure or tree vulnerable to ignition.[33] Installation can also be challenging during extreme weather events like high winds or temperatures.

Installation of the blankets can also damage the structure or tree in which it is trying to protect. For example, tears in the blanket can trap water in between the structure or tree and the blanket, leading to water damage. In addition, the staples and screws needed for installation can compromise or damage the structure or tree.

While aluminized blankets are touted as a cost-effective technique for protecting one's assets, they may be out of reach for many homeowners (to cover an average single-story home, one would need around $2500).[34] This may create an equity issue in fire-prone areas like the WUI. In addition, it is simply not feasible for federal agencies to purchase enough blankets to save every historical structure or heritage tree.

In the future, it might be worthwhile to consider built-in, automated aluminized systems for structural protection. Additionally, other materials beyond aluminum, silica, fiberglass, and Kevlar could be explored.[35]

Lastly, while aluminized blankets may offer some protection for structures and heritage trees in fire-prone areas, the technique could just be considered a technological fix. Perhaps the false security of techniques like these might actually promote more development in hazardous areas and reckless behavior, de-incentivizing the use of other wildfire risk reduction techniques.

1 mi

Healdsburg

03

*Site Hardening*

Russian River

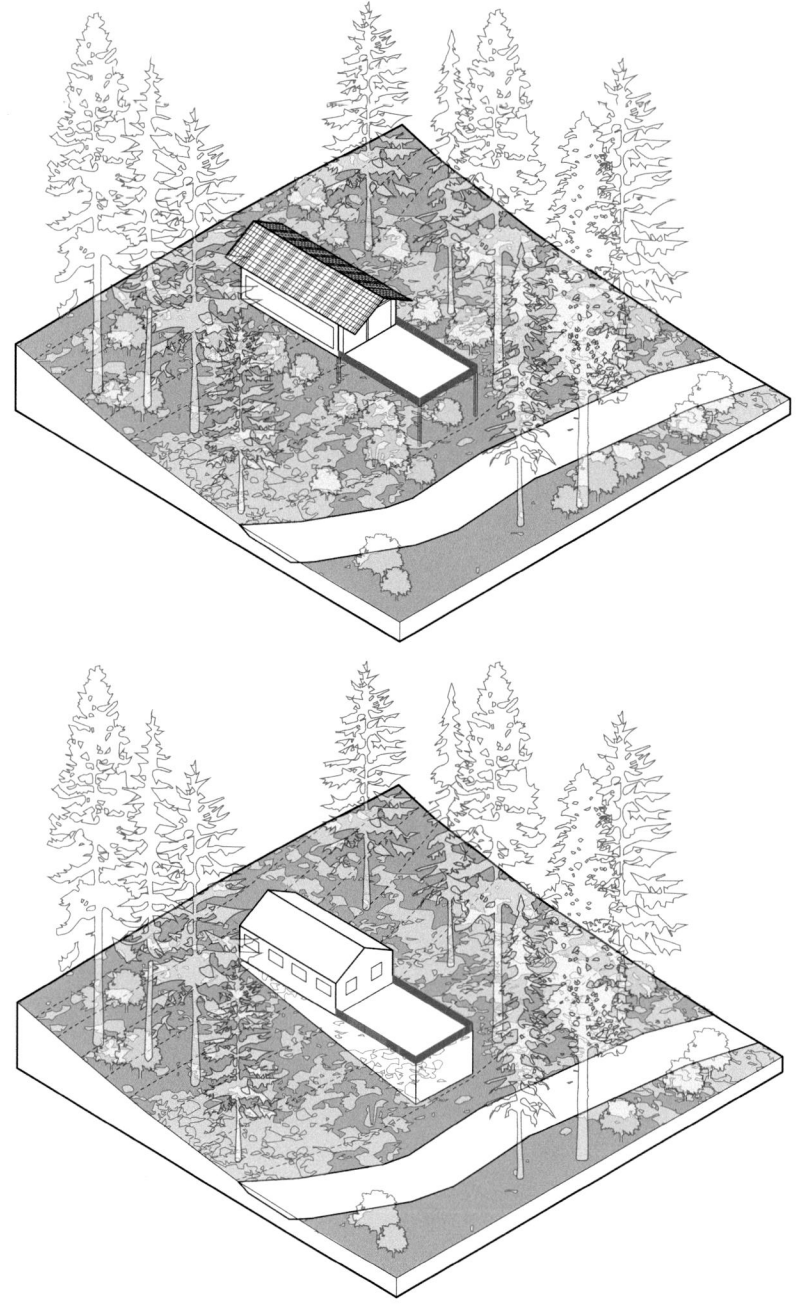

03

*Site Hardening*

| | |
|---|---|
| Description | Retrofitting existing sites to be more resistant to ignition |
| Example Location | Healdsburg, California |
| Size | Variable |
| Primary Implementer | Homeowner |
| Stakeholders and Team Members | N/A |

## Technique Overview

Typically, the initial ignition of a structure happens through spotting, and through radiant heat transfer when buildings are preheated from nearby vegetation or structures that are burning. While these ignitions tend to start as small fires, they can quickly grow, sometimes leading to structure-to-structure fires which can impact entire neighborhoods.[36]

Protecting structures in the WUI is expensive; approximately 50–95% of all wildfire suppression costs in the U.S. is solely dedicated to this. In an effort to reduce this burden on taxpayers, bolster community resilience, and increase owner responsibility in reducing property-level risk, several standards have been developed. While some of these standards are mandatory and enforceable by law, most just serve as voluntary guidelines to help landowners in the WUI better adapt to the growing threat of wildfire.

The strategies embedded in these documents tend to focus on three primary ways to reduce property-level ignition risk. First, there is the creation of defensible space in the home ignition zone by maintaining vegetation around structures. There is also the careful selection of building materials. Lastly, there is the strategy of designing the building envelope so that exterior materials fit together in a way that reduces ignition risk.[37] This case study focuses on the last two strategies: material selection and building envelope design. Many times, these techniques are lumped together under the concept of site hardening.

## Planning and Design Process

Figure 4.11 (*Previous*) Aerial of Healdsburg. Source: Google Earth 2022.

Figure 4.12 (*Left*) A view of a residence prior to wildfire hardening (top) and the same view with various retrofits to make the site more resistant to wildfire damage (bottom).

Site hardening often begins with the roof. Highly combustible materials like wood are discouraged for roofs; instead, materials like tile, metal, and asphalt are recommended. Additionally, complex roof architecture may create more spaces for combustible debris like branches and leaves or embers to collect; therefore, it is recommended that roof design be as simple as possible. It is also suggested that owners routinely maintain their roofs to cover new gaps, prevent critters from building nests, and remove any collected debris.[38] Lastly, the connection between the roof and other parts of the structure is important; metal flashing is, therefore, recommended at all edges to create a physical buffer.[39]

Drainage gutters located at the edges of roofs should be made of metal, have guards installed to prevent the collection of debris, and be routinely cleaned especially ahead of fire season. Alternatively, roof drainage could be rerouted internally with subsurface outlets. The space under roof overhangs is also an area of concern as this area can trap heat and <u>embers</u>. To reduce risk, building owners can reduce the lengths of overhangs or enclose exposed eaves with a <u>noncombustible</u> material.[40]

Intentional gaps that connect the interior of a structure to the outside are also vulnerable in wildfires due to <u>ember</u> intrusion. Intentional gaps can include rooftop vents for fans, stovetops, dryers, and HVAC units as well as chimneys.[41] To prevent embers from entering these gaps and igniting a structure, it is recommended that owners install fine mesh at all thresholds.[42] Additionally, siding can channel flames upward and gaps in siding can pull flames inside a structure. To prevent this from happening, it is suggested that all flammable siding be replaced with <u>noncombustible</u> material, that siding start at least a 6″ above the adjacent ground, and that any damaged siding be replaced.[43]

Windows and skylights can easily break during a wildfire event creating vulnerable gaps in the building envelope. Most standards recommend replacing single-pane and wooden- or plastic-framed windows with multi-paned, metal-paned, tempered-glass windows. All windows should have screens installed to prevent ember intrusion. Also, if possible, the number and size of windows facing wildland vegetation and steep slopes should be reduced.[44] Furthermore, all exterior doors, including garages, should be made of a noncombustible material, have weather stripping, and can be manually operated.

Some standards also recommend whole-house treatments to reduce ignition risk. These can be automated systems like heat-detecting roof-mounted sprinklers that douse the structure and adjacent vegetation in water.[45] Property owners can also manually apply coatings like gels or foams ahead of a wildfire.

Decks and fences should also be evaluated, especially if they are attached to or are in close proximity to a building.[46] Ideally, decks should be made of noncombustible materials, have flashing installed at joints, be completely enclosed, and be surrounded by noncombustible materials (no close vegetation). Additionally, it is advisable to put decks on flat ground, not on a slope.[47] Fences should not be attached to buildings. If they are, the last five feet closest to the building should be made out of a noncombustible material. Additionally, fence gaps should be closed to prevent the collection of debris and embers and nearby vegetation should be cleared.[48]

While many site <u>hardening</u> examples exist across the world, the work of engineer Chris Arai in Healdsburg, California is especially remarkable. In the early 2000s, Arai began hardening his ridgeline property. To do this, Arai implemented a number of the strategies outlined earlier in this case study. For example, his buildings had simple rooflines made of metal with short overhangs. Above-grade, the exterior walls were made of lime plaster; below-grade, they were one foot wide and made with Durisol and filled with concrete and steel. He installed a whole-house sprinkler system and connected it to his pool; when activated, this system doused the site with water. To ensure the system worked during emergency shut-offs, he installed a series of solar panels with battery storage. Additionally, before evacuating, he applied hydrogel around all of his windows.[49] When the 2019 Kincade Fire swept through his neighborhood, the structures on his property were the only ones left standing. In total, the fire burned over 400 structures, including 200 homes.[50]

## Performance and Evaluation

Studies have shown that hardening techniques can effectively reduce wildfire-related damage to structures. Furthermore, while a comprehensive hardening project may seem daunting and expensive to property owners, site retrofitting can happen incrementally over time with smaller projects. To help prioritize action items, vulnerability assessments can be employed. Also, most retrofitting projects do not require specialized construction skills, so most builders and contractors could be used for implementation.[51]

Beyond wildfire risk reduction, many of these techniques reduce long-term maintenance and extend material lifespans. For example, asphalt roofs, fiber cement siding, and plastic composite decking all require less maintenance and last longer than traditional, more combustible materials. These techniques can also increase energy efficiency. For instance, replacing single-pane windows with multi-pane windows can significantly improve the energy efficiency of a structure.[52]

## Challenges and Future Research

While site hardening techniques reduce the structural ignition potential for existing buildings, it is impossible to fireproof. For a structure to ignite, it only needs one weak link, like a mesh-less vent or a combustible attached fence. Additionally, some site elements cannot be modified; for instance, if a house is built upslope of prevailing winds, hardening techniques may not matter. Thus, site hardening techniques can give property owners a false sense of security, prevent people from evacuating during a wildfire event, and can even promote more development in hazardous areas.

Also, even though retrofitting projects can happen in stages, it is still an expensive process and household income is a key variable for implementation. For instance, replacing a roof – the most vulnerable part of a structure – is one of the most expensive upgrades.

Furthermore, even though some retrofits may reduce long-term maintenance, property owners still need to be vigilant about inspecting and maintaining their property to ensure efficiency.[53] For older structures, this is even more important.

Another challenge for site hardening is the need for collective action. This is especially important for denser, fire-prone areas with many property owners. Even if one resident implements a range of ignition reduction techniques, it may not matter if their neighbors do not.[54]

Site hardening codes are also imperfect. Many policies just apply to new construction, allowing older homes to get grandfathered in. Also, only a few states (like California) have adopted wildfire-related building codes, and even with these codes, enforcement is expensive and logistically challenging.

Lastly, as with many other techniques, many property owners may not be motivated to implement site hardening until they have personally experienced a wildfire event.[55]

1 mi

Whittlesea

Kinglake National Park

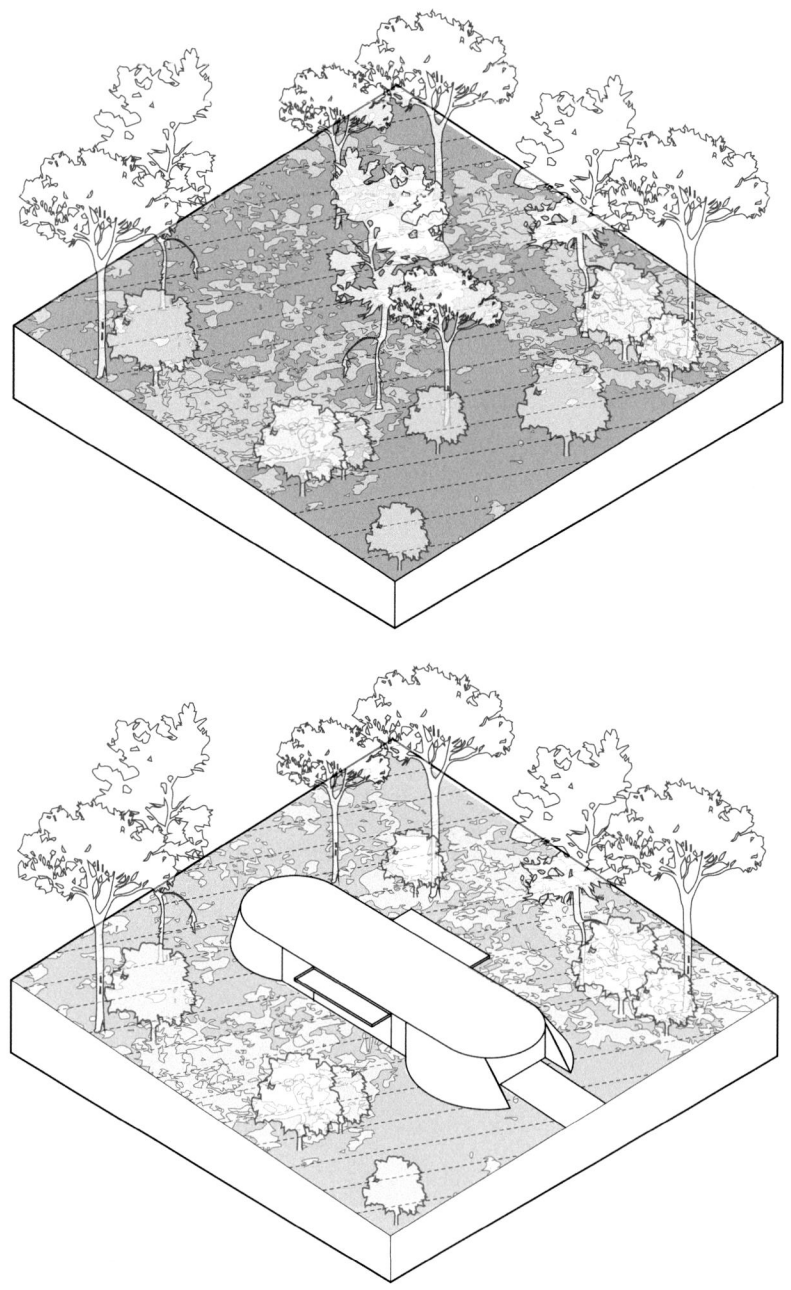

04

## *Home Bunkering*

| Description | Designing new "fire-proof" earth-sheltered structures |
|---|---|
| Example Location | Whittlesea, Australia |
| Size | ~ 5000 sf |
| Primary Implementer | Baldwin O'Bryan Architects |
| Stakeholders and Team Members | Property Owners |

## Technique Overview

On February 7, 2009, a series of fires erupted across the Australian state of Victoria and, over the course of a month, over 1,000,000 acres burned, 2,000 structures were destroyed, and over 170 lives were lost. Following the fire, Australia's wildfire evacuation policy came under scrutiny; unlike other fire-prone countries, Australia endorses a "prepare, stay, and defend or leave early policy." Residents can opt to stay on their property and actively protect their assets from fire or they can evacuate well ahead of a fire front. Unfortunately, during the "Black Saturday" bushfires, nearly 70% of the fatalities occurred as a result of people sheltering in a building.[56]

In the wake of this tragedy, a small group of architects convened as part of the "We Will Rebuild" initiative to brainstorm ways in which residents could build back smarter by adopting new approaches for wildfire-safe home construction.[57] Their rationale was that conventionally-built houses burn easily from ember attacks due to the combustible materials used in construction, and their traditional architectural features. Furthermore, they recognized that when a fire reaches an area with multiple structures, it can shift from being a wildfire to an urban fire, with conventional houses fueling the flames.[58]

In general, three main principles guided design decisions for the "We Will Rebuild" initiative. New houses needed to be able to survive a fire front, exclude embers from entering the structure, and prevent structure-to-structure spread. Additionally, if someone plans to shelter in a new structure during a fire event, they should be able to safely remain inside for 30 minutes if the structure is far from other structures or 60 minutes if it is near to other structures.[59]

Figure 4.13 (*Previous*) Aerial of Whittlesea. Source: Google Earth 2022.

Figure 4.14 (*Left*) A view of a landscape without any development (top) and the same view with a now wildfire-safe home designed to withstand conflagrations (bottom).

## Planning and Design Process

Following the 2009 bushfires in Victoria, architect Sean O'Bryan with Baldwin O'Bryan Architects started receiving a slew of requests to design fireproof homes from residents who had either lost their homes or nearly lost their homes. O'Bryan's architectural practice is unique in that they focus on residential designs that can better withstand the threat of oncoming fires. Their designs have five defining features.[60]

First, their homes are sheltered by the surrounding ground. In some situations, the homes are built into the sides of slopes. In other situations, the architects peel back the surface of the ground, construct the homes, cover the structures with a thick layer of soil, and then regrade the landscape so that it slopes back to the existing grade.[61] Another defining feature of these homes is that they are primarily made of compressed earth blocks which are noncombustible; the soil functions as a buffer between the interior of the home and the flames and the conditions outside. The primary structure of these homes are prefabricated steel arches that bolt together. The homes also have minimal openings that connect the interior to the exterior to allow for as much coverage as possible. To protect these openings during a bushfire and fully seal the homes, mechanized fire curtains or roller shutters are installed.[62]

In 2015, the firm won the Bushfire Building Council of Australia's Innovation Award for their conceptual framework, and in 2020, they completed construction on a fire-resistant home called the Whittlesea House.[63]

## Performance and Evaluation

New fireproof earth-sheltered homes have obvious benefits. First, and foremost, they offer some protection from bushfires and, thus, peace of mind for developers, residents, and owners living in fire-prone areas. Some even suggest that the unique design features of these homes might actually encourage residents to safely evacuate ahead of a fire front, instead of staying on the property to defend it. The protection these homes provide stems from the primary material used in construction: earth.[64] Compressed earth blocks have excellent thermal qualities; they are heat-resistant and they create a large buffer to insulate the interior from the exterior. Furthermore, some claim that fires can actually strengthen compressed earth block construction as the heat physically bakes the structure.[65] The added protection from this kind of construction also comes from the deliberate limiting of interior-exterior connections and potential ember entry points. These homes do not have eaves, gutters, valleys, ridges, gables, or gaps in roofing material – all key areas for embers to get caught during a fire event.[66]

Beyond their fire-resistant qualities, these homes are relatively fast to construct. This is, in part, due to the prefabricated steel frames that form the basic structure of the homes. Using this prefabricated, bolted system significantly reduces labor time when compared to conventional construction. According to Baldwin, the structural components of the walls and roof of a typical home can be erected in just one and a half days. This savings in labor also corresponds to a savings in cost.[67] Furthermore, the compressed earth blocks can be made from local soil, saving costs on materials and transportation. These homes also tend to use less energy in heating and cooling (especially with passive solar design) and use less water than traditional homes.[68]

These homes have other benefits as well. They can provide a larger carbon sink for the property and create habitat for onsite plants and animals. Some also suggest that this architectural style better blends into the site landscape and can be adapted to different types of sites. Due to their sheltering properties, they can also have better acoustics than conventional buildings. Lastly, some fire protection measures have multiple functions; for instance, roller structures can help prevent glare and insects from entering the home.[69]

## Challenges and Future Research

While newly built structures designed to withstand <u>bushfires</u> help to neutralize fear, there is no guarantee that a structure will not burn due to the unpredictability and variation of current bushfire events. The idea of "fire-proofing" is simply misguided. In the case of the 2009 Black Saturday fires, for example, 20% of those who died in their homes were deemed well-prepared to <u>shelter-in-place</u>.[70] Furthermore, academic research on fireproofing houses is thin, with most research coming from those working in construction (who have a vested interest in proving its benefits).[71]

Beyond a lack of evidence for bushfire survival, these kinds of homes do not use traditional materials or follow typical construction methods.[72] Thus, it may be difficult to find contractors willing to build them. Also, while these homes may have some cost savings for owners looking to rebuild, the expense may still be cost-prohibitive for many residents in fire-prone areas. This issue points to a larger concern about shifting fire safety responsibility to homeowners, leading to a slew of private solutions that just widen the equity gap. Some even consider this to be a form of disaster capitalism. Furthermore, after a bushfire event happens, many neighborhoods are labeled either "successful" or "unsuccessful" in resisting fire in order to receive public funds. Yet, "successful" neighborhoods tend to be more affluent and able to afford private solutions, putting them at a clear advantage over other places.[73]

There is also a concern that the general public may not be open to this kind of home construction due to the desire for more conventional architecture. Some people still equate earth-sheltered homes with "bomb shelters" – claiming that they are dark, damp, and lack good airflow.[74]

Along with other bushfire technological fixes, fireproof homes may provide a false sense of security for residents and may actually promote unsafe practices in fire-prone areas. To start, the trend of building fireproof structures may catalyze development in hazardous areas, leading to more sprawl. This not only puts more property and lives at risk but could actually increase fire risk due to human ignition. The trend may also promote unsafe sheltering-in-place practices. If residents believe that their homes are actually fireproof, they may choose to ride out a fire at home. But as we know from the 2009 bushfires, this does not always go as planned.

One alternative to consider for this technique is the construction of an onsite personal bunker. Instead of fireproofing an entire home, could residents purchase a small in-ground structure that could be used as a last resort in the case that evacuation is not possible? These types of bunkers rose in popularity following the 2009 Black Saturday <u>bushfires</u>. Often marketed as "somewhere to go when there's nowhere to go," they are made out of concrete and are partly submerged in an earthen mound. The entrance to the bunker is often made of heavy-gauge stainless steel and faces away from the prevailing wind direction. Inside, there are typically lights, an air circulation system, a thermometer, a phone charger, breathing masks, and a survival kit. The bunker can serve as a refuge for around 30 minutes, just enough time for a typical <u>fire front</u> to pass through a property.[75]

1 mi          Table Mountain National Park                    Kenilworth

*Firebreak Cutting*

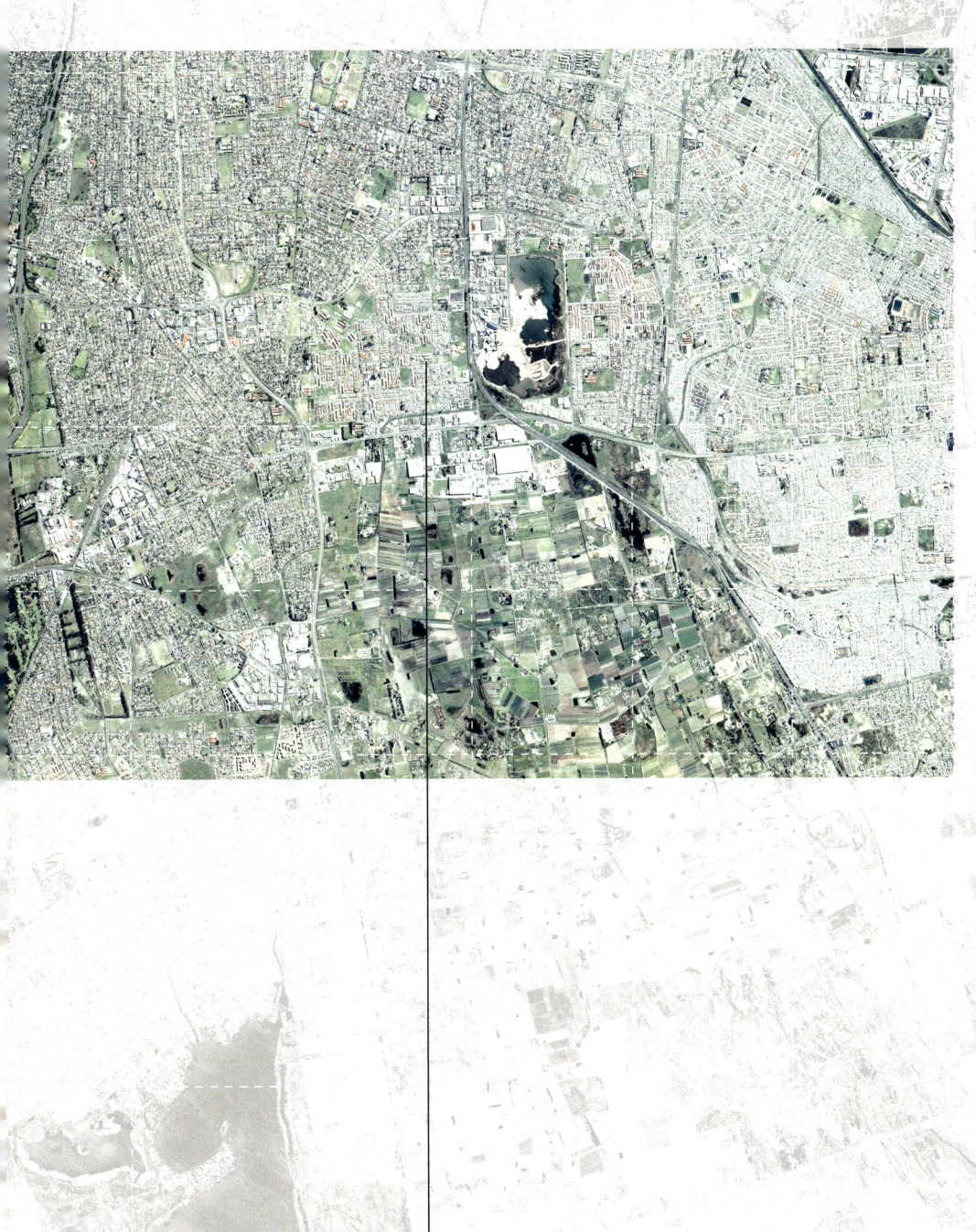

Hanover Park

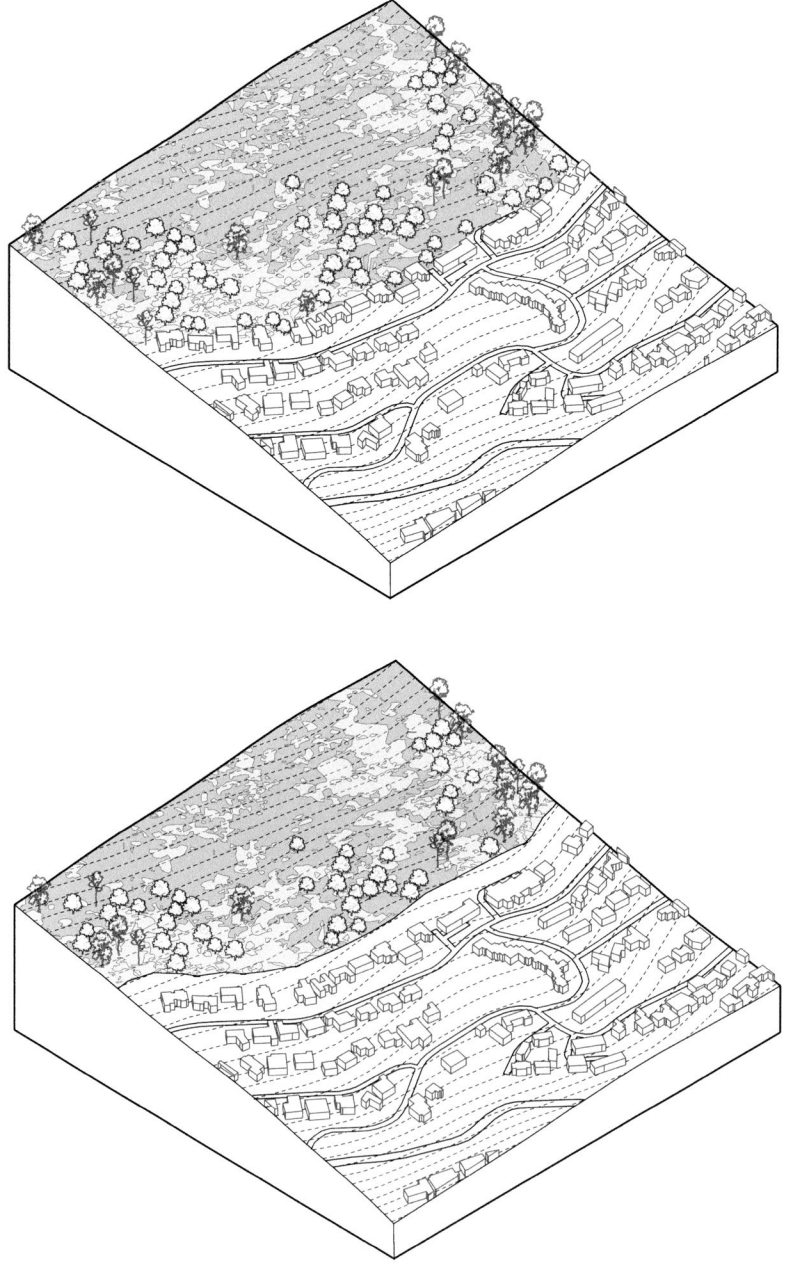

05

*Firebreak Cutting*

Description          Cutting vegetation in linear strips across the landscape
Example Location     Cape Peninsula, South Africa
Size                 120,000 acres
Implementer          Ukuvuka
Team                 City of Cape Town, SANParks, Cape Peninsula Fire Protection Committee

## Technique Overview

The Cape Peninsula of South Africa is a fire-prone and fire-adapted region where urban development is encroaching into the wildlands of nearby hills and mountains. Prior to 1998, wildfire protection, suppression, and management activity in the area was scarce, uneven, and disjointed given the large number of landowners and public land managers with jurisdiction in the region. Then, in 1998, the government founded the Cape Peninsula National Park and established the Forest Fire Act which increased and unified wildfire management initiatives in the region. Following a particularly destructive fire in 2000, a campaign called Ukuvuka arose. In just a few years, Ukuvuka implemented a number of wildfire risk reduction strategies including a circum-peninsula firebreak system.[76]

A firebreak is a strip of land, either constructed or natural, that serves as a line of defense between wildland and the built environment. Constructed firebreaks are linear segments of landscape, which are stripped of vegetation. Natural firebreaks can be roads, waterways, or rock formations that are devoid of flammable materials.

Today, the City of Cape Town, the South African National Parks organization (SANParks) and several local Fire Protection Committees (FPAs) help to maintain the circum-peninsula firebreak network.[77] Three management areas comprise this 86-mile-long network, extending from Cape Town in the north to Cape Point in the south.

## Planning and Design Process

There are four main methods that have been used in constructing the circum-peninsula firebreak: burning, ploughing and grading, herbicide application, and hoeing, brush cutting and mowing.[78]

Historically, Cape Peninsula firebreaks were built using a prescribed burning method. With this process, managers constructed two to three firebreaks next to one another and rotationally burned them so that at least one of the breaks was devoid of vegetation.[79] While burning can be cost-effective for creating smaller firebreaks, it can have several drawbacks: this process can lead to runaway fires, putting nearby assets at risk; safe burn days can be limited due to unpredictable weather patterns; and the method can require professional expertise.[80]

Ploughing and grading is another technique used for firebreak construction. While tractors and graders can cover a large area in a short amount of time with this method, the price of fuel and equipment rentals can be costly. Furthermore, this technique only works on flat land; if implemented on a slope, soil erosion could be problematic.[81]

Herbicide application is another option, especially in areas with large concentrations of invasive alien species (IAS). This is an expensive and labor-intensive method, though, and can negatively affect indigenous plants.[82]

Of the four firebreak construction methods, hoeing, brush cutting, and mowing is the preferred technique on the Cape Peninsula. This method can be used in most weather conditions, except when the fire danger index (FDI) is orange or red, and can be repeated annually. It is also considered more environmentally-friendly than the other options and requires less skilled labor.[83]

Beyond the method by which the firebreak is constructed, there are also several other design considerations. The first is the width of the firebreak. Typically, firebreaks in this region are 30 to 60 feet wide, but this could vary based on the height of adjacent vegetation (the higher the vegetation, the wider the break). Firebreaks are not typically built on steep slopes to prevent erosion issues and because firefighters may have a difficult time using these areas during an event. It is recommended that, instead, firebreaks are constructed along contours or on ridges. Plant regrowth is another consideration as some areas, like south-facing slopes, regrow quickly and need more maintenance. All fuel cut from a firebreak needs to be evenly dispersed at least 60 feet away from the strip or removed from the site entirely. Furthermore, firebreaks tend to be oriented parallel to the prevailing wind direction during fire season to prevent embers from the fire front from easily blowing over the break. Sensitive plants or habitat should be avoided or transplanted. Lastly, those building firebreaks should incorporate existing natural or built features as much as possible. For example, one could brush cut along a road verge, creating a wider and less intrusive firebreak.[84]

## Performance and Evaluation

Firebreaks like those constructed in the Cape's circum-peninsula network can help to slow or prevent the spread of fire from wildlands to built areas with high-value assets by serving as a linear barrier of reduced fuel load. This is especially true if the firebreaks are constructed using the design considerations outlined above and if wildfire events are small and low-intensity.

Firebreaks can also increase access for firefighters to reach wildfire events and to provide a space for suppression activities like conducting a direct attack or lighting a backburn.[85]

Lastly, the wildfire management program that emerged in the late 1990s created a much more unified, streamlined, and strategic approach for the Cape Peninsula compared to the disparate initiatives that came before. It also helped to build capacity in frontline communities with increased wildfire risks by offering employment opportunities related to localized fuel management.[86]

## Challenges and Future Research

Even if firebreaks are constructed with proper considerations for width, topography, plant regrowth, fuel disposal, maintenance, and prevailing winds, they sometimes cannot slow or stop an oncoming fire. Some wildfires, especially those that are high-intensity and wind-driven, will spot across firebreaks regardless of their design. Thus, this technique is often implemented in tandem with other wildfire risk reduction measures.[87]

Also, while the construction of firebreaks using hoeing, brush cutting and mowing has low ignition potential when compared to other methods, there is still a concern of equipment unintentionally throwing sparks or overheating which can lead to fires. In addition, in some situations, firebreaks may encourage unwanted access to adjacent neighborhoods and landscapes. This could increase human-induced ignition risk.

Firebreaks can also have visual impacts on the landscape and alter regional viewsheds. This is especially true when they run perpendicular to slopes. In these conditions, firebreaks can also instigate and exacerbate erosion issues.[88]

In some situations, the construction of a firebreak may require written authorization from the government if there is potential for a significant environmental impact on the landscape. This may increase project costs and extend the timeline for completion of the activity.[89]

Lastly, while the City of Cape Town, SANParks, and several local FPAs help to maintain the circum-peninsula firebreak network, these entities have limited resources. Thus, it is necessary for adjacent landowners, both public and private, to share the responsibility of designing, constructing, and maintaining the breaks.[90] Across a large region, this can be difficult to implement.

It should also be noted that firebreaks are just one type of wildfire "break" that can be implemented to slow or prevent the spread of fire. While firebreaks eliminate almost all flammable material from a strip of land, other types of breaks, like fuelbreaks function as a fuel reduction zone. In this space, some vegetation is allowed to grow which can maintain habitat continuity. Some breaks, called shaded fuelbreaks can even incorporate tree canopy as long as the trees are appropriately spaced and maintained.

1 mi

Tuolumne River

*Ridge Clearing*

Groveland

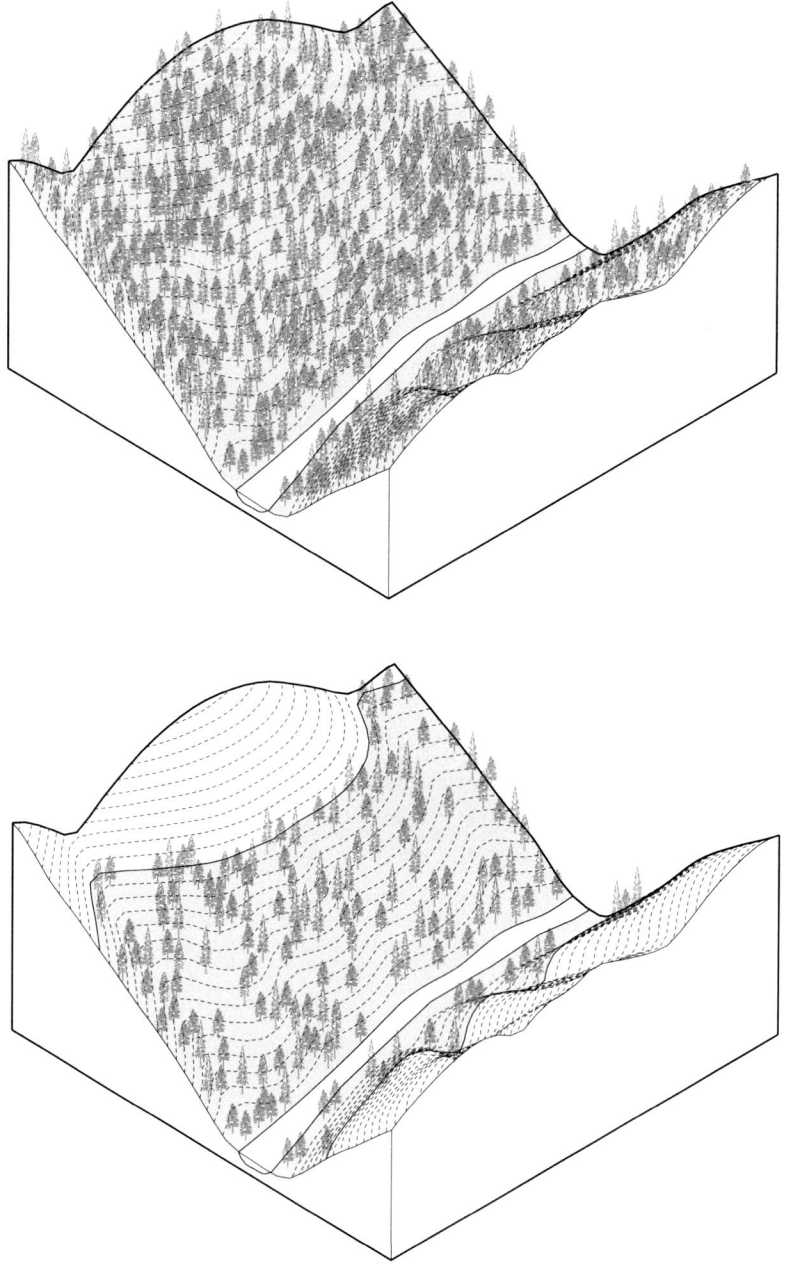

06

*Ridge Clearing*

| | |
|---|---|
| Description | Clearing vegetation on key ridgelines to create "fire plains" |
| Example Location | Tuolumne County, CA |
| Size | ~500,000 acres |
| Primary Implementer | Hanna Prinssen (conceptual proposal) |
| Stakeholders and Team Members | Yttje Feddes, Joyce van den Berg, and Gert-Jan Wisse |

## Technique Overview

The history of wildfire in Tuolumne County, California closely resembles the history of wildfire within the larger Sierra Nevada landscape. Prior to Euro-American settlement, indigenous populations tended the landscape for generations, using fire as a key maintenance tool. The mosaic-like habitats that were created, and the species that were fostered as a result of these habitats, helped to sustain an indigenous way of life and economy.

In the 1800s, when colonists arrived, they mistakenly thought these tended landscapes were "natural" and implemented policies to suppress fire on the landscape. Today, as a result of the fuel buildup arising from this wildfire suppression, coupled with a warming climate, and an expansion of the wildland-urban interface (WUI), the risk of catastrophic wildfire in places like Tuolumne County is very high.[91]

In 2020, Hanna Prinssen developed a Master of Landscape Architecture thesis project addressing these topics at the Academy of Architecture, Amsterdam. She did this after attempting to visit Yosemite National Park in 2018, but was thwarted by the Furgeson Fire. During this particular wildfire season, over 8,500 fires burned over 1.8 million acres across the state. Her resulting thesis, entitled "A Fire-scape: A new form of a fire resilient landscape," focused on how to design with fire instead of bolstering defense lines against it in the western foothills of the Sierra Nevada mountain range in Tuolumne County. This part of California features a gradient of vegetation, from cultivated landscapes in the lower elevations of the Central Valley, to oak grasslands in the foothills, and dense conifer forests further upslope. It is also a landscape that attracts a wide range of tourists and visitors given its proximity to Yosemite National Park.[92]

Figure 4.17 (*Previous*) Aerial of Tuolumne County. Source: Google Earth 2022.

Figure 4.18 (*Left*) A view of a highly-vegetated landscape (top) and the same view with cleared ridgelines and a central "fire plan" that is allowed to burn (bottom).

## Planning and Design Process

At the cornerstone of the design proposal is a toolbox of strategies that could be employed in Tuolumne County. These strategies challenge the conventional relationship we have with fire by allowing some wildfires to burn, using fire to catalyze economically productive landscapes, and foregrounding $CO_2$ storage as a primary goal. The strategies were informed by a close reading of the existing landscape including the hydrology, forest structure, fire regime, development patterns, and recreational opportunities of the region.

One prominent strategy in this toolbox is the clearing of vegetation along key ridgelines that run perpendicular to prevailing winds during fire season. These linear strips are created by excavating to granite bedrock; because without soil, vegetation cannot grow and become <u>fuel</u> for fire. Also, since fire tends to move faster uphill, these cleared areas serve to slow down or stop the flames before they reach key areas of inhabitation. These areas are considered safe zones and function as a buffer for those living near to the ridgeline.

In between these cleared ridges is a valley landscape that is designed to burn. This is another strategy in the toolbox. In this "fire plain," wildfires are encouraged to burn to promote the creation of a mosaic-like forest that has a range of habitats and supports a range of species. This zone can also be used as a production forest with the north-facing slopes primarily growing softwood and the south-facing slopes primarily growing hardwood. These trees can then be harvested, producing income for local communities. Over time, this forest can also alter future <u>fire behavior</u> by diversifying the existing forest.

At the lowest point of the "fire plain" is an area where natural streams flow. In order to slow down the water, dampen the surrounding landscape, and encourage infiltration, weirs are designed in the existing channel. This pooling of water also creates new wet meadow landscapes, increasing local biodiversity.[93]

## Performance and Evaluation

Ridge clearing may help to slow or stop wildfires from escaping valleys due to the significant <u>fuel reduction</u> efforts involved. This technique can also create a <u>staging area</u> and help firefighting personnel appropriately position themselves in the case that a fire in the "fire plain" burns too quickly or too severely.

Additionally, by creating a safe space for fire to burn, the project could help reveal the ecological, social, and economic benefits of fire on the landscape to residents and visitors, which could, in turn, help shift long-standing views about wildfire being a destructive element in the landscape.

Lastly, by diversifying the forest composition through increased management, the project can create habitat heterogeneity and boost biodiversity in the region. This diversification, along with the deliberate cultivation of certain species, could help create a production forest as a revenue generator. The outputs from this process could help to economically sustain the landscape management process.

## Challenges and Future Research

While ridge clearing is a promising strategy, it should be noted that <u>firebreaks</u> like these do not always succeed at slowing or stopping <u>fire fronts</u>, especially for wildfires that are high-intensity or are <u>wind-driven</u> with <u>embers</u>. Thus, care should be taken to appropriately consider elements like width and ongoing maintenance regimes.

This kind of technique is also a labor-intensive and expensive process to clear the landscape down to bedrock at such a large scale. These cuts in the landscape may also negatively impact viewsheds and could accelerate erosion issues.

Lastly, there are so many landowners and stakeholders in Tuolumne County that would need to have a shared set of goals to greenlight this kind of project. This group includes a range of WUI residents, timber production companies like Sierra Pacific Industries, electrical companies, and public entities like the United States Forest Service which oversees the Stanislaus National Forest.

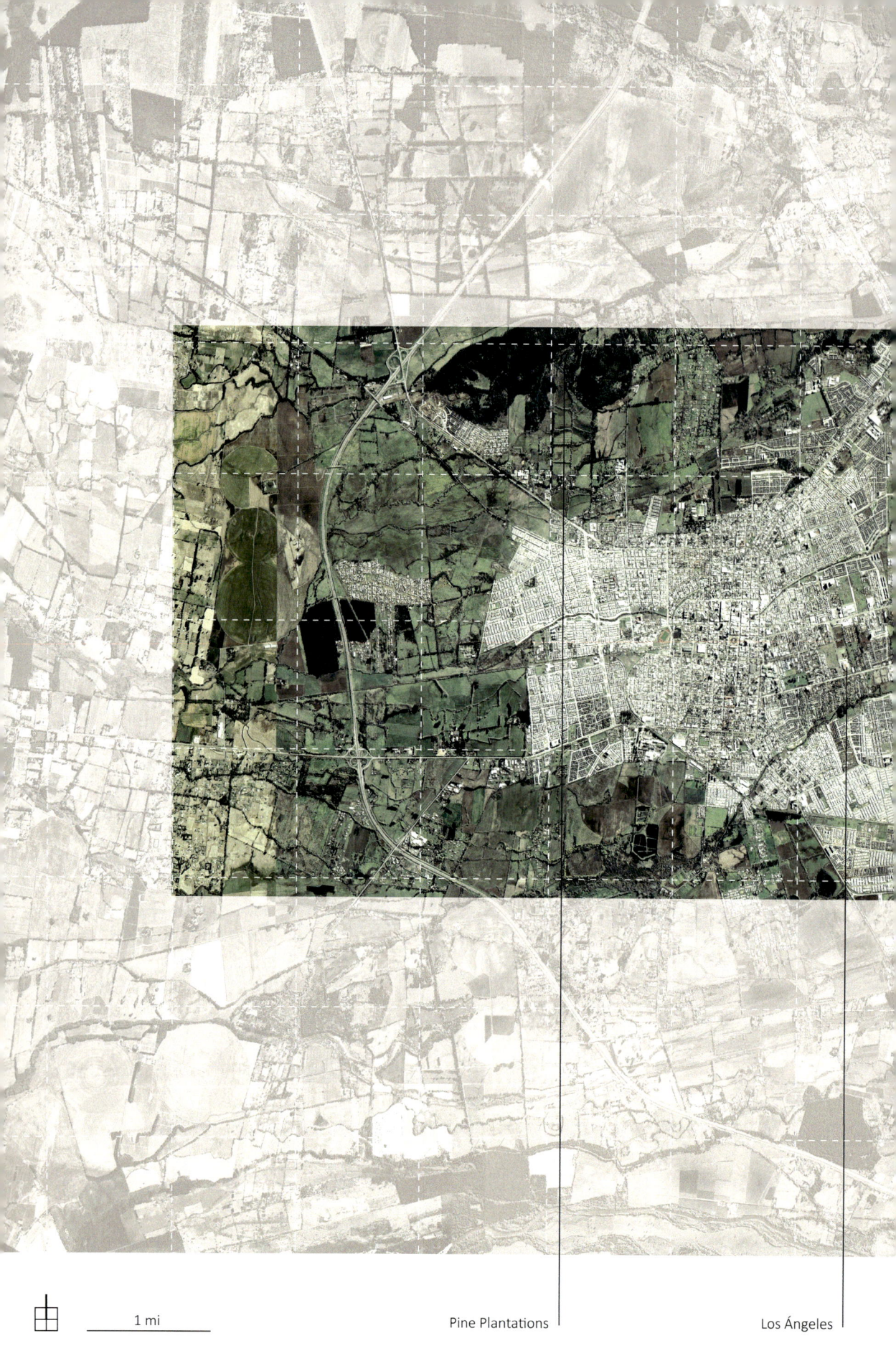

1 mi

Pine Plantations

Los Ángeles

*Donut Extracting*

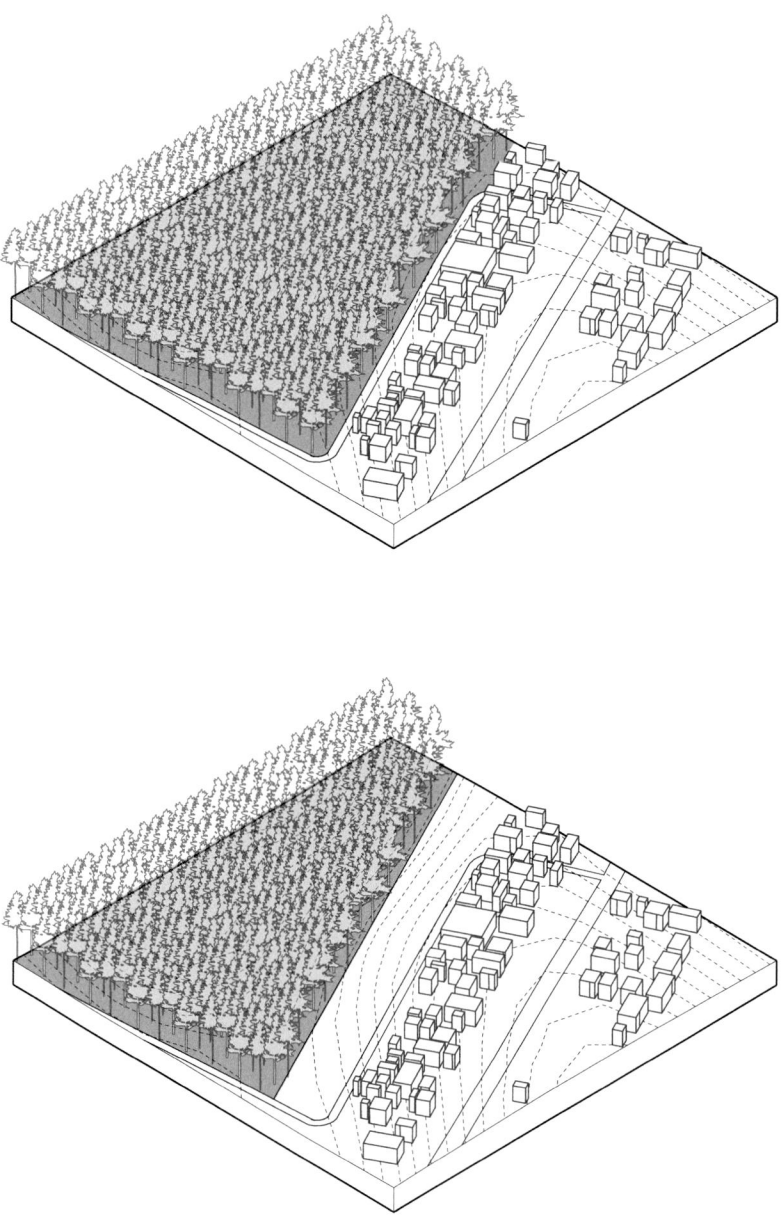

07

*Donut Extracting*

| | |
|---|---|
| Description | Creating a citywide fire protection buffer by removing all flammable material |
| Example Location | Los Ángeles, Chile |
| Size | 17,500 acres |
| Implementer | Regional Government of Biobío |
| Team | National Forestry Corporation (CONAF), Ministry of Agriculture, Municipality of Los Angeles, forestry companies, electricity distribution companies, road administration companies, electrical maintenance companies, and local residents/landholders |

## Technique Overview

The community of Los Ángeles is located in Central Chile between the rivers of La Laja and Biobío. It sits within a wide valley tucked between the Andes Mountains to the east and the coastal mountains to the west.[94] Economically, this region of Chile contains 44% of the country's forest plantations and is a key exporter of wood products for paper plants and mills. Similar to the rest of Chile, these plantations have replaced natural forests and former grazing land, significantly altering land cover across the country.[95] This land cover shift has contributed to an increase in wildfire risk due to the high fuel load and flammability of the plantations, along with the plantations being a key target for arson. A shift in development pressure has also contributed to regional wildfire risk as more residents move into interface areas outside of city centers.[96] In Los Ángeles, as with the rest of Chile, most forest fires are caused by human activity, as either accidental ignitions or arson.[97]

Following a particularly destructive wildfire season in 2015, many Chilean researchers and wildfire professionals began supporting landscape management activities and community design techniques to limit the environmental, social, and economic impacts of future wildfire events.[98] Then in 2016, the National Forestry Corporation of Chile (CONAF) designed a Forest Fire Prevention program for the region, focused on 24 highly vulnerable communities. This program, which was funded by the Regional Government of Biobío, had six main goals: educate communities about the tangible steps they can take to reduce the risk of ignition, understand the causes behind recent wildfire events, identify high-risk areas, build capacity in both the public and private sectors, train residents in fire prevention measures, and implement strategically-located fuel reduction projects.[99]

Figure 4.19 (*Previous*) Aerial of Los Ángeles. Source: Google Earth 2022.

Figure 4.20 (*Left*) A view of a neighborhood built directly next to a pine plantation (top) and the same view with a firebreak intended to protect the community (bottom).

## Planning and Design Process

The last goal of the Forest Fire Prevention program, the implementation of fuel reduction activities, focused on preventative forestry projects in the interface zone of the 24 identified communities. In general, the interface is where forest fuels about urban development – with either a hard, defined edge or a fuzzy, undefined boundary. In the Los Ángeles plan, the team defined this zone by offsetting current municipal boundaries approximately 1.25 miles, creating a citywide buffer zone for potential interventions.[100]

After geographically defining the potential buffer zone, the team then developed a protocol for prioritizing areas within the Los Ángeles' interface for fuel reduction activities. After conducting a literature review, surveying local forest fire experts, and consulting with local residents, they overlaid six spatial datasets: a heat map of recent forest fires, nearby population centers, regional road networks, population densities, vegetation typologies, and landscape slopes. With this information, they identified three priority areas in the interface: one near Villa los Professors, one near Villa Genesis, and one near Villa las Tranqueras. The team deemed these areas as having the highest risk of forest fire in Los Ángeles.

In these interface areas, the team constructed firebreaks to break up the continuity of the vegetation, change the behavior of oncoming fires, and to protect community assets. To do this, they used heavy machinery like bulldozers to remove existing vegetation and to expose mineral soil. The firebreaks they created formed continuous strips at the edges of these three areas in the interface.[101]

While only 12 miles of firebreaks have been constructed to date, the goal is to extend the technique around the edge of Los Ángeles, creating a donut-like citywide protective barrier.[102]

## Performance and Evaluation

As with other firebreak-focused projects, this initiative in Los Ángeles has the potential to slow or stop an oncoming wildfire from catastrophically affecting community infrastructure located in the interface. By removing all flammable material down to mineral soil, the strips break up the existing masses of vegetation that serve as fuel during a wildfire. There is a significant amount of gray literature that supports this claim.

Furthermore, this firebreak zone can be used as a community refuge space or a staging area for wildfire professionals during an active wildfire event. Community members can retreat to this space during an emergency or firefighters can use this space for direct or indirect attacks.

Also, the use of heavy machinery like bulldozers, to implement this technique can reduce costs and save time when compared to more labor-intensive methods.

## Challenges and Future Research

While a citywide fire protection buffer, like the one being implemented in Los Ángeles, has the potential to reduce the risk of fire spread, it may not be able to protect against wind-driven fires or high-intensity wildfire events. Even if the buffer is wide and void of flammable material, some wildfires cannot be slowed or stopped.

Logistically, there are many challenges with implementing and maintaining a citywide firebreak. In the interface of Los Ángeles, local residents privately own much of the land; thus, there is a need for extensive community outreach, coordination, and buy-in.

There is also the immense scale of the project that requires a stable, long-term funding source for continued construction and maintenance. Heavy machinery used for clearing could also unintentionally throw sparks into the adjacent landscape or overheat, increasing the risk of wildfire.

Firebreaks, especially those that expose mineral soil, can also negatively influence viewsheds, increase erosion potential, and accelerate habitat loss. They can also create unproductive dead spaces at the periphery of development. While more flammable, wider fuelbreaks and shaded fuelbreaks could be explored as viable alternatives.

1 mi

*Defensive Spacing*

Kranshoek                                                          Plettenberg Bay

08

*Defensive Spacing*

| | |
|---|---|
| Description | Removing fuel in <u>home ignition zones</u> and protecting vulnerable communities |
| Example Location | Kranshoek, South Africa |
| Size | 3,000 acres |
| Implementer | LANDWORKS (formerly Kishugu NPC) |
| Team | Individual property owners, United Nations Development Programme (UNDP), Working on Fire (WOF), National Fire Protection Association (NFPA), Global Environment Facility (GEF), Republic of South Africa Department of Environmental Affairs (DEA), local municipality, South African Police Service (SAPS), traditional leaders, local Fire Protection Associations (FPA), South African National Parks (SANParks), Cape Nature, U.S. Forest Service (USFS) |

## Technique Overview

In 2002, the NFPA worked with the U.S. Forest Service (USFS), the U.S. Department of the Interior (DOI), and state forestry groups to develop a program called <u>Firewise</u> USA for reducing the loss of lives and property due to wildfire by making vulnerable communities more physically resilient and by educating residents about the environmental drivers that put them at risk.[103]

Two years after the <u>Firewise</u> USA launch, a non-profit organization called LANDWORKS, based out of Cape Town, began collaborating with NFPA to translate and adapt these principles for communities in South Africa.[104] This process led to the development of four models that could be applied to a range of communities in the Western and Eastern Cape: (1) a model based on voluntary efforts, intended for more affluent communities, (2) a semi-voluntary model that involved trained residents receiving wages for part-time work, (3) a model focused on community employment and was often used in higher poverty areas with fewer work opportunities and (4) a model focused on self-sustaining strategies for communities interested in funding their own initiatives. With this last model, community members created businesses that supported wildfire risk reduction work.

Then, in 2012, LANDWORKS received funding to jumpstart a community-based wildfire management program called FynbosFire; it began with four pilot communities located in the Cape: Kranshoek, Sir Lowry's Pass Village, Clarkson, and Goedverwacht.[105] This case study will focus on the work being implemented in Kranshoek.

## Planning and Design Process

In Kranshoek, located between Knysna and Plettenberg Bay, a small group of residents employed through the FynbosFire program conducted a number of tasks aimed at reducing

Figure 4.21 (*Previous*) Aerial of Kranshoek. Source: Google Earth 2022.

Figure 4.22 (*Left*) A view of a house with significant vegetation and combustible material around it (top) and the same view with fuel removed in key ignition zones (bottom).

the risk of wildfire events in their community. These tasks included: measuring and communicating fire danger, raising residential awareness about defensible space, educating school-aged children, clearing overgrown vegetation in public places, and setting up longer-term community safety protocols.

Twice a day, the FynbosFire team measured the fire danger index (FDI) by recording the outside temperature and relative humidity and accounting for wind and rain. If the potential for burning was extreme, they flew a red flag over the community; if it was high, a yellow flag; if it was moderate, a green flag; and if it was low, a blue flag. The flag was used as a communication tool for the larger community – indicating when residents need to be extra careful about ignition sources and debris around their homes.[106]

The team also spent a significant amount of time going door-to-door in the community to raise awareness about residential fire safety. In these conversations, they focused on defensible space in the home ignition zone – how to create buffers around each home by clearing garbage, overgrown vegetation, and anything flammable. During these sessions, they also promoted Firewise gardens, which consist of fire-resistant plants like succulents and are arranged with horizontal and vertical breaks. In certain situations, like with elderly residents, team members helped to create or maintain defensible space.[107] Educating the younger residents of Kranshoek was also a priority for the FynbosFire team. To do this, they spent time at local schools, raising awareness about FDI flag messaging, defensible space strategies, and evacuation plans. The hope was that these young people would bring fire safety knowledge home with them.

A fourth task of the FynbosFire team was to help clear and maintain public spaces around Kranshoek. One area that was of particular interest was under elevated power lines; here the team removed overgrown vegetation that could ignite. They also did this along roadways in town as well as in ditches at the edge of the community. They also focused on areas of the community where wood is harvested to make sure they are well-maintained.

To ensure longer-term community safety for Kranshoek, the FynbosFire team ensured that all houses were numbered, all street names were labeled, and all hydrants were visible and accessible. Additionally, they created a step-by-step guide on how to evacuate during a natural disaster. This information was posted up in publicly accessible areas of the community. Lastly, the team updated the community's Fire Management Plan and disseminated this document to stakeholders and residents.[108]

## Performance and Evaluation

The FynbosFire program empowers communities to take ownership over their future and to develop initiatives for the greater good. Through this program, the communities see themselves as problem-solvers, able to develop local solutions to help reduce the impacts of wildfires on their communities.[109] Many of the residents in these communities also realize that the program is about much more than just fire. FynbosFire is also about building community and capacity for a more resilient future – goals which have helped with retention in the program.[110]

FynbosFire also supports knowledge sharing at multiple levels. At a neighborhood level, there is the ability for peer-to-peer exchange, where residents can share their tips and strategies with one another. At the community level, there are opportunities to train-the-trainer, and to allow those people to share their expertise through workshops. Then, at the regional level, there is potential for sharing findings across communities through social networking, newsletters, and conferences.[111]

There is also evidence that the FynbosFire program shifts people's perceptions and behaviors about fire risk reduction. While the effects of the program have yet to be systematically measured, there is anecdotal evidence that the program has reduced damage to property and the loss of lives. For example, Kranshoek was spared from the 2017 Knysna Fires despite being in the line of fire – many residents of the community partially attribute its survival to the risk reduction activities of FynbosFire.[112]

FynbosFire also improves the economic conditions of communities like Kranshoek. It does this primarily through job creation and through skill development.[113] Through the program, participants can learn a range of skills from financial management to the operation of hand tools, and can even start their own business.[114] Additionally, the program improves the condition of the landscape in and around participating communities, which can positively impact local farming operations and other landscape-based businesses.

Lastly, the FynbosFire program is adaptable, scalable, and replicable; it can function in a range of communities with varying social, economic, and environmental conditions and can be modified in the same community through the annual renewal process.[115]

## Challenges and Future Research

One significant hurdle for the FynbosFire program relates to communication – how to clearly convey why residents are at risk and why they should participate in the program to reduce their risk, which is critical for the community risk assessment. Ideally, this assessment should be a report with easily understandable text and graphics.[116]

Another challenge is the voluntary nature of the program as extensive cooperation between residents is necessary. Yet, oftentimes, residents have different, competing agendas. For instance, most houses have overlapping home ignition zones, so developing a singular risk reduction strategy between homeowners can be difficult. Furthermore, it is often hard for community members to agree on the management of certain public spaces. For example, in some public spaces, wattle is grown and harvested for fuel, but is also considered a wildfire nuisance.[117]

Participation can also be tedious. It involves a lot of paperwork and many meetings to build trust and develop collective goals. After action items have been implemented, it also takes a lot of effort to systematically measure the success of outcomes to guide future work. Lastly, when sharing experiences and findings with other communities, it is often necessary to translate as South Africa has 11 official languages.[118]

The last significant challenge relates to retention – how to sustain participation in the program, especially once supportive funding is no longer available. For many vulnerable communities, it is difficult to continue the fire risk reduction efforts without stipends. And while it is possible, and encouraged, for communities to develop self-sustaining enterprises to fund the work, it takes time and expertise to develop successful business plans.[119]

## *Defensive Spacing Interview*

Figure 4.23 Val Charlton, Managing Director at LANDWORKS.

DESIGN BY FIRE: Could you provide some background about LANDWORKS and your role in the organization?

VAL CHARLTON: The beginning goes back to January 2000 which was when we experienced a large fire in Cape Town and immediately after it was under control, a project started called Ukuvuka Operation Firestop. That's when I started Working on Fire (WOF) because I thought it was a really worthwhile project. It was a four-year campaign in Cape Town focused on the natural-urban interface. It was a public-private partnership between an insurance company, a media company, the City of Cape Town, and what was the Department of Water Affairs and Forestry.

And at the same time, new legislation around forest fires called The National Veld and Forest Fire Act was published. So, Ukuvuka helped to develop this new act at a landscape level. Toward the end of that campaign, people began realizing that these initiatives could happen nationally, which is how the WOF program started. And at that stage, the WOF program was run under a non-profit company that eventually became LANDWORKS.

The WOF program eventually got so big, that the mission of the non-profit shifted toward community education, which is how LANDWORKS came about. At that point, we started picking up on Firewise principles. I made contact with people doing this work in the U.S., including the Forest Service and the National Fire Protection Association (NFPA) and we started the roll-out of Firewise. And we've been working with NFPA ever since.

DF: Could you speak to the importance of community education around wildfire risk reduction principles?

VC: Here at LANDWORKS, we think knowledge is power so we place a lot of emphasis on educating people about Firewise and what it can do for them, but it's never that simple. For example, in one rural community outside of the Western Cape, we had done door-to-door outreach for ten years and were still having issues connecting with residents. And at a close-out workshop in that community, a community leader said "Well, you still go to church on Sunday, don't you?" And I thought, "That's true! You actually do keep going to church on Sunday even though you know what the message is."

In the Western Cape, there tends to be more affluence. And we noticed that if a community is affluent, they often believe that they are entitled to fire protection. They pay their taxes and believe it is someone else's job to protect them. They also are more likely to have fire insurance which would pay them out in the event of a fire. Having said this, if you ask people on the urban edge about Firewise concepts like "defensible space" and "ember attacks," most are familiar. Whether they do anything about it is another story.

DF: I'm curious about what you see on the horizon for LANDWORKS? What would you like to be focusing on in the future?

VC: I don't even know how to answer that now because the last few years have been extraordinarily difficult for us to keep things running and government funding has essentially dried up. Perhaps if we have a few difficult fire seasons in the near future, people might sit up and start paying attention. But my dream is that in five years, the focus is less on LANDWORKS as an organization, and that the Firewise principles we support become a national program, either voluntary or paid. That's what we would like to see.

1 mi

Angeles National Forest

Glendora

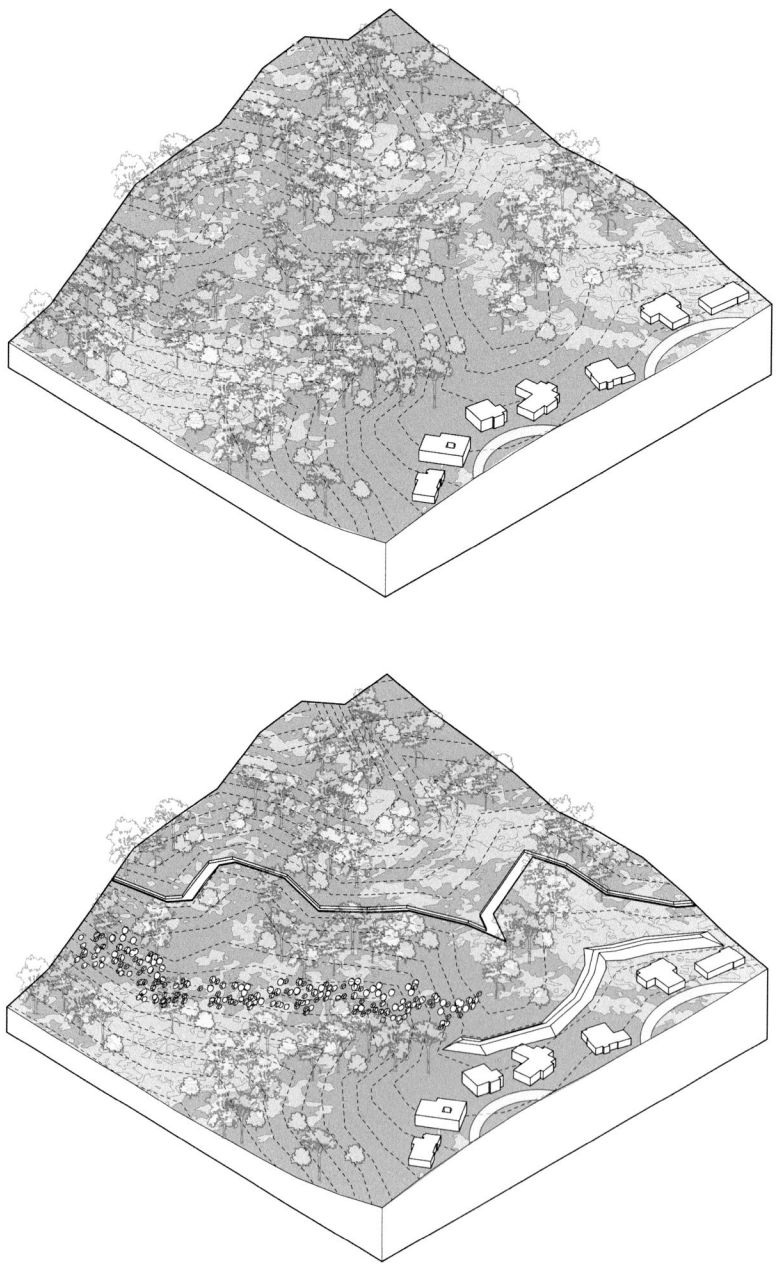

09

*Earth Shifting*

| | |
|---|---|
| Description | Excavating, depositing, and berming soil as <u>firebreaks</u> |
| Example Location | Glendora, CA |
| Size | Variable |
| Implementer | Sarah Toth (conceptual proposal) |
| Team | Catherine Seavitt Nordenson and Matthew Seibert |

## Technique Overview

The foothills of the Angeles National Forest in Southern California are highly vulnerable to catastrophic wildfires. In this area, sprawling development, primarily in the form of single-family residences, abuts uninhabited vegetated landscapes. To provide protection for at-risk communities in the WUI who sit within or adjacent to these fire-prone foothills, local and regional fire agencies have long used a range of fire suppression techniques. Additionally, they have begun employing <u>prescribed burning</u> and <u>mechanical thinning</u> techniques to further reduce risk. While these two techniques are often successful at reducing wildfire risk in alpine and sub-alpine forests, they have the potential to catalyze landscape conversions in the chaparral and sage shrublands found in this area, opening the landscape up to invasive grasses. These fast-growing grasses can eventually replace the dense shrubs, becoming more <u>fuel</u> for future fires.[120]

In 2018, Sarah Toth, a Landscape Architecture masters student at The City College of New York, developed a design project, "Pyro-Diversion: Planning for Fire in the San Gabriel Valley," that challenged this blanket and non-site specific approach.[121] Instead of continuing to use these general techniques, she proposed a new set of strategies – developed strategically for the Glendora landscape. The goals of these strategies were three-fold: (1) to slow down and redirect <u>fire fronts</u> in the fire-prone foothills, (2) to minimize threats to nearby private properties, and (3) to still allow for wildfires to burn the fire-adapted landscape. The techniques work in tandem with existing fire infrastructure already found in the foothills, like debris basins and fire roads, and are intended to function as a flexible, multi-pronged approach to fire management.[122]

Figure 4.24 (*Previous*) Aerial of Glendora. Source: Google Earth 2022.

Figure 4.25 (*Left*) A view of a neighborhood adjacent to an uninhabited vegetated landscape (top) and the same view with three earth shifting strategies to reduce the impact of wildfire (bottom).

## Planning and Design Process

In this set of strategies for Glendora, three focus on the concept of earth shifting in designated watersheds to change <u>fire behavior</u>. The other three strategies involve revegetating slopes dominated by flashy, non-native grasses, planting shelterbelts perpendicular to the dominant wind direction to slow wind and trap <u>embers</u>, and promoting aluminum blanket house shields for vulnerable structures adjacent to the wildlands to deflect embers, flames, and heat.

The first earth shifting strategy involves creating 25-foot-wide linear firebreaks by excavating down to granite bedrock. In this part of the foothills, bedrock can be found just 3' below the surface. Once excavated, this soil can be transported and reused further downslope. Then, throughout the rainy season, these linear swaths can fill up with water, creating new habitat and supporting local species. This strategy can also be used to protect specific parts of the landscape from burning.

The second strategy involves gathering large stone aggregate from existing debris basins in the area and transporting them to create 25-foot and 50-foot-wide firebreaks running perpendicular to the dominant slope direction. These stones sit atop a non-woven geotextile and are planted with fire-resistant native vegetation with high moisture levels. During fire events, these swaths of stone aggregate can help to slow down the fire.

The third earth shifting strategy involves gathering finer sediment from nearby debris basins to build large, 25-foot-high berms along the edges of the foothills, defining a new development boundary and curbing future construction. These berms can then set the stage for future interventions. The sediment piles can be capped, supported from below, and vegetated on the north-facing side. Over time, the berms can replace high-risk residences that have been bought out, with residences relocated further downslope in less vulnerable locations.[123]

## Performance and Evaluation

Developing site-specific wildfire adaptation strategies for California is necessary as each location has a unique set of conditions that cannot be addressed with blanket solutions. In this case, the three earth shifting strategies function as linear firebreaks in the landscape that have the potential to slow down or stop fires and serve as staging areas for firefighting crews. These techniques can also give residents who live nearby more time to evacuate ahead of a fire front.

Furthermore, the technique involves repurposing waste material from the series of existing debris basins. Typically, this material would otherwise go unused.

Lastly, the techniques may help with erosion control issues in the foothills and may decrease water velocity as it travels downslope. These are important considerations as wildfire-affected landscapes can be prone to mudslides and debris flows in the rainy seasons following fire events. Unfortunately, these slides and flows can be catastrophic to adjacent ecosystems and neighborhoods.

## Challenges and Future Research

Despite these positive aspects of earth shifting, the technique faces a number of implementation challenges. First, sifting, transporting, and depositing a large amount of sediment and aggregate is an expensive and labor-intensive process that involves heavy machinery in difficult landscape conditions.

Also, if installed incorrectly without proper stabilization protocols in place, the firebreaks may actually instigate or exacerbate erosion issues in the foothills. This could be especially problematic in the wake of a wildfire event when slopes are void of plant material.

Lastly, it may be difficult to convince residents and local stakeholders to sign off on the plan for a range of reasons. The technique may substantially impact local viewsheds with firebreaks along hillsides often being very visible. Allowing wildfires to burn so close to homes in the WUI may also raise some red flags for nearby residents. Lastly, the notion of relocation – either voluntary or mandatory – is still a politically fraught and largely unsupported idea by the general public.

1 mi

chaparral and sage scrub shrubland

Ramona

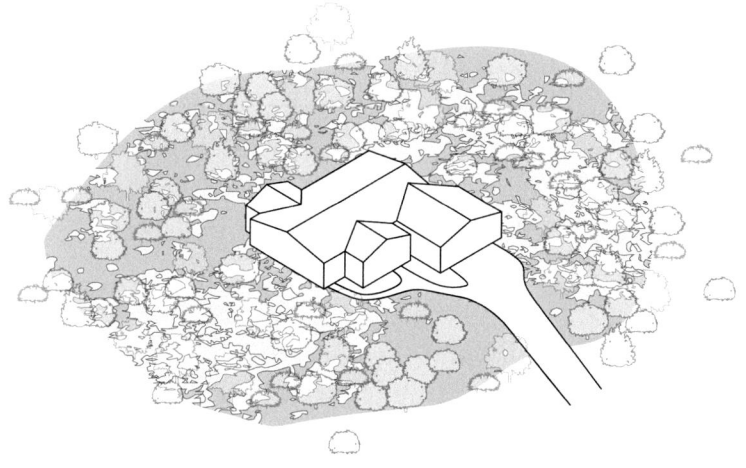

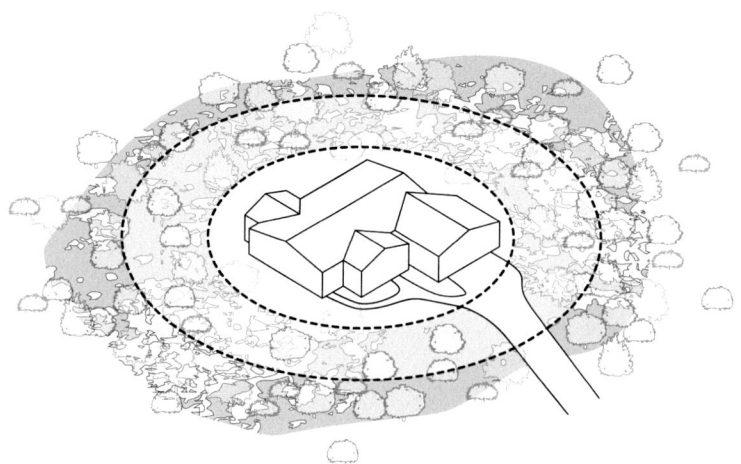

# 10

## Ring Tending

| | |
|---|---|
| Description | Creating green belts of lightly irrigated native vegetation around structures |
| Example Location | Ramona, CA |
| Size | Variable |
| Implementer | U.S. Geological Survey, Western Ecological Research Center, Sequoia-Kings Canyon Field Station |
| Team | Department of Ecology and Evolutionary Biology at UCLA, California's Own Native Landscape Design, Inc. |

## Technique Overview

In the chaparral and sage scrub shrubland in San Diego County, development is growing deeper and deeper into wildland. Here, native plants are considered a major fire risk due to their live fuel moisture content (LFMC), which measures how much water is within a plant. In this area, native plants have between 60% and 80% LFMC; meaning that they are susceptible to ignition, can spread flames to nearby plants, and can increase flame lengths.

To help manage this risk, the clearing of vegetation around structures in fire-prone areas, also known as creating defensible space, is a widely accepted practice. Yet, many homeowners are reluctant to do this because they like having plants and trees close to their homes for reasons such as increased privacy, aesthetics, and energy efficiency. As a result, many homeowners resort to leaving their yards unmaintained or planting non-native species around their structures as protective greenbelts for wildfire risk reduction.[124]

In 2020, a group of researchers from the USGS Western Ecological Research Center, the Department of Ecology and Evolutionary Biology at UCLA, and California's Own Native Landscape Design, Inc. sought to challenge these common practices in San Diego County. To do this, they set up a study to understand if native plants could be managed in a way to reduce wildfire risk around homes.[125]

## Planning and Design Process

Their study focused on three residential landscaping projects near Ramona, CA in San Diego County's wildland-urban interface. The three projects were at least five years old at the start of data collection, and all three had a similar landscape structure. Closest to each home, about 30 feet from the edge of the structure, was a ring of native shrubs that had been lightly irrigated during the dry season, from June to October. Beyond that first ring, between 30 feet and 100 feet from the house, were native shrubs that had been recently thinned, and beyond the thinned area, more than 100 feet from the house, were native shrubs that had been left untreated. All three sites were on relatively flat ground.[126]

The primary hypothesis developed by the team was that the first ring comprising lightly irrigated native shrubs closest to the house would provide the most protection during a wildfire event by limiting flame length and fire spread. To test this hypothesis, the team monitored the LFMC of the plants every two weeks within each of the three rings. They collected data for two and a half years to better understand seasonal changes. Then, they modeled the fire behavior of the three zones to assess the efficacy of the different treatments.[127]

## Performance and Evaluation

As they anticipated, the LFMC for the plants in all three rings reached 100% during the winter rainy season. Yet, throughout the rest of the year, especially in summer and fall when precipitation was limited, the LFMC varied significantly in each of the different treatments being observed. In the summer months, the lightly irrigated plant ring had the highest LFMC levels, followed by the thinned ring, and finally the untreated ring.

These LFMC findings, along with the fire behavior modeling outcomes revealed that the lightly irrigated zone was the most successful in reducing flame length and the rate of fire spread around the homes. The authors also suggested that this green vegetation could help catch and extinguish embers ahead of a fire front. Thus, if homeowners are reluctant to thin or clear vegetation around their structures through common defensible space practices or plant non-native species in a greenbelt, this approach gives them another option.[128]

Additionally, lightly irrigating native shrubs around one's home can improve overall aesthetics of the property, increase property values, and help to maintain privacy. It can also create valuable habitat for native fauna like pollinator insects that rely on local plants, while increasing the biodiversity of other species.

Lastly, when compared to traditional non-native greenbelt practices, this approach tends to have lower water needs. This could be an important consideration for areas that are susceptible to drought.[129]

## Challenges and Future Research

While the idea of cultivating a lightly irrigated native shrub ring around one's home has a lot of potential in reducing residential wildfire risk, there are some challenges.

First, this study is limited in data as it only compares three sites, and as the authors suggest, most other evidence about this practice is anecdotal. Thus, it would be helpful to perform a larger study in the future to see if these initial findings still hold true with more data.

Additionally, the performance of these yard designs could be compared with those that feature non-native plants, as well as yards that have been largely cleared of vegetation. Doing so would provide a more holistic assessment for homeowners who are weighing different landscape management options for their homes.

Future studies could also consider different spatial arrangements of vegetation as well as other environmental aspects like varied slopes and wind directions, as all of these elements play a role in how fire moves across a site.

1 mi

## Agricultural Stripping

Reserva Natural Llaberia

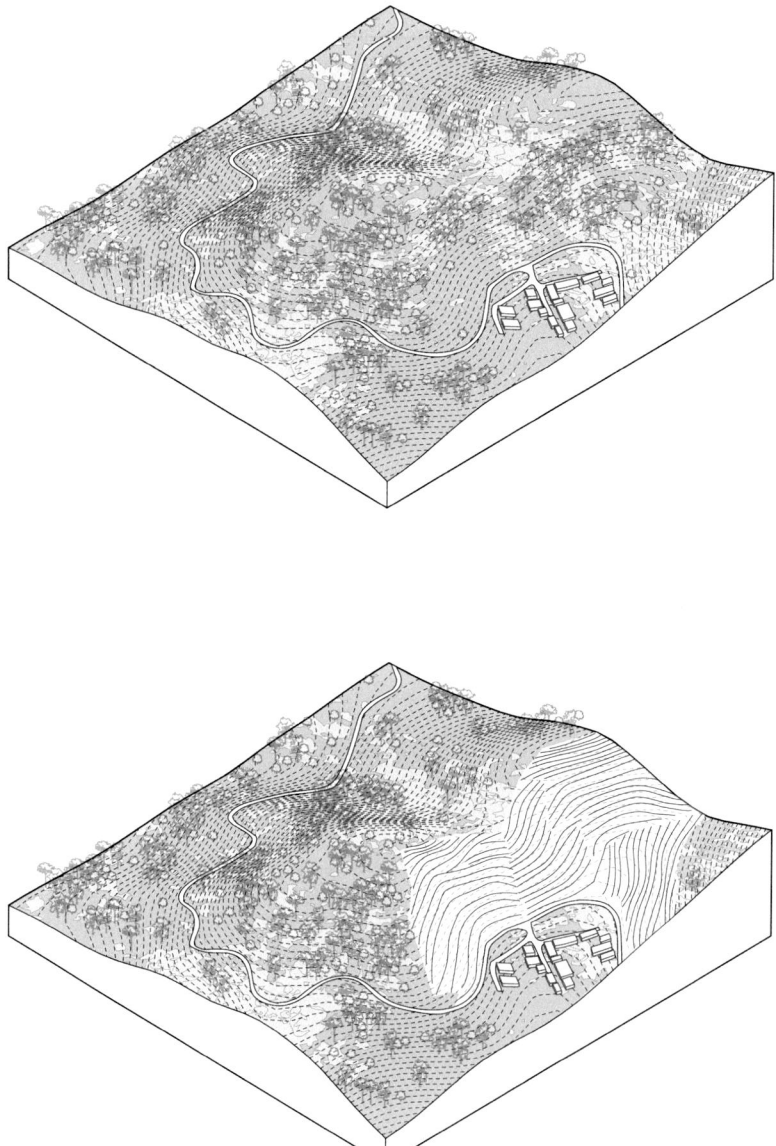

# 11

## *Agricultural Stripping*

| | |
|---|---|
| Description | Promoting vineyards and truffle plantations as protective perimeters |
| Example Location | Tivissa-Vandellòs-Llaberia-Pradell, Spain |
| Size | 150,000 acres |
| Primary Implementer | Ministry of Agriculture of Catalonia |
| Team | General Directorate of Forest Ecosystems and Environment Management, Department of Agriculture, Livestock, Fisheries and Food, local municipalities, landowners, Forest Defense Groups, Catalan Fire Service, Consorci of the Serra de Llaberia |

## Technique Overview

In the Mediterranean basin, <u>priority protection perimeters</u> (PPPs) often dictate regional forest planning and management. PPPs are densely forested areas that have a high risk of being impacted by large (<1,200 acre) wildfires and are typically bounded by infrastructure like highways. Each PPP has its own wildfire prevention plan based on existing management capacity and landscape qualities like vegetation composition and quality and local meteorological conditions. Each plan lists a series of recommended interventions to change the behavior of wildfire so that it slows, stops, or loses intensity.[130] The goal of these wildfire prevention plans is to outline a series of actions that can form an integrated and holistic vision for reducing fire risk in the PPP.

In the Catalonian region of Spain, 34 PPPs collectively cover over 2,500,000 acres.[131] One of these PPPs, named ET2 Tivissa-Vandellòs-Llaberia-Pradell, sits in a mountainous region of Catalonia with a high wildfire risk. This area has been poorly managed and has seen a proliferation of highly flammable plant species as a result. About 76% of the PPP comprise coniferous forests, primarily Aleppo pine and associated species. The area also has a high natural value, as it contains seven sites included in the national Plan for Spaces of Natural Interest (PEIN).[132]

Figure 4.28 (*Previous*) Aerial of Catalonia. Source: Google Earth 2022.

Figure 4.29 (*Left*) A view of a mountainous village tucked within the landscape (top) and the same view with a planted vineyard to act as an edible buffer between the community and wildlands (bottom).

## Planning and Design Process

In the late 2000s, a multidisciplinary group of experts got together to draft a wildfire prevention plan for ET2 Tivissa-Vandellòs-Llaberia-Pradell. The process for developing a PPP plan typically consists of three main steps. The first step focuses on collecting, inventorying, and mapping existing data about the PPP and included topics such as: historical <u>fire perimeters</u>, projected wildfire risk, meteorological information, roads, water points, <u>fuelbreaks</u>, <u>firebreaks</u>, landfills, power lines, housing, protected habitat, and areas of cultural significance.

The second step involves analyzing the above information to understand potential threats. The final step is a proposal for action that includes a range of projects for the PPP.[133]

In 2010, the group completed these three steps and published a plan focused on four main goals: to enlarge low-fuel zones, to improve road access, to maintain water reserve points, and to create strategic firefighting control points. The group listed the first goal as their highest priority, with 84% of all actions in the plan supporting the reduction of fuel in key areas of the PPP.[134]

One unique action listed in this plan was the protection and creation of edible fire buffers, like vineyards and truffle plantations, in strategic areas of the region. In areas where production was already happening and was helping to serve as a buffer between wildland and the built environment, the group of experts suggested enlarging the area and reinforcing the operation with additional water points. They also identified uncultivated areas that could benefit from fire prevention greenbelts and suggested that these zones could be used for future cropland.[135]

## Performance and Evaluation

Anecdotally, there is evidence across many Mediterranean regions that agricultural greenbelts like vineyards can slow, stop, or reduce the intensity of an oncoming wildfire. In some cases, the first few rows of the greenbelt burn but the rest of the crop survives. In other cases, only the cover crop burns. There are a few reasons for this. First, the layout and composition of an agricultural greenbelt play a role in its efficacy as a fire buffer. For instance, if a greenbelt is tightly planted and consists of a single crop, it may have a better chance of buffering flames. Second, if the greenbelt has a high water content, this could also be beneficial in the face of an oncoming wildfire. This high moisture content can come from specific plant qualities but could also come from irrigation as a part of cropland management.[136]

Agricultural stripping can also improve firefighting operations in the event of a wildfire. This space can serve as a staging area for professionals to mount direct or indirect attacks through firebreak construction, backfire ignition, or dropping retardant or water. In many cases, agricultural areas have access to water that can be used for suppression purposes. For instance, frost protection sprinklers in vineyards can be turned on ahead of a wildfire to douse nearby land and infrastructure.[137]

There are also many economic benefits to implementing this technique. First, since the technique is profitable, there is a chance that private landowners would be willing to pay for implementation, creating a self-sustaining fire buffer. This technique also bolsters rural development by providing training and jobs for the local community. It also creates a value-added product that could increase tourism for the region. For instance, many wineries in Monstant are marketing their vineyards as wildfire prevention zones.[138]

There are also additional benefits like carbon sequestration, increased land value, and creating more ecosystem services. Working lands like vineyards and plantations can capture carbon from the atmosphere through proper soil and habitat management protocols.[139] They can also increase land values by improving views and making fallow land productive. Lastly, there is the potential to manage croplands in a way that increases ecosystem services.[140]

## Challenges and Future Research

Unfortunately, agricultural stripping is not fireproof. Any buffer or break, regardless of its flammability or width, cannot routinely change the course of an approaching wildfire. This is especially true with <u>wind-driven fires</u> that blow <u>embers</u> miles ahead of the flame front or high intensity fires that emerge from extreme weather conditions.[141] Some consider agricultural greenbelts only to be "micro-breaks" with the ability to protect nearby structures.[142] Furthermore, while significant anecdotal evidence exists about the efficacy of agricultural buffers, there are few empirical studies that quantify their impacts.

Initial funding and management is another hurdle. While this technique has the potential to be privately-funded, most pilot projects are funded through subsidies or public investment. Furthermore, in order to be successful, stakeholders involved in the project need to share the same vision and agree on a cooperative management structure.[143]

Water usage is another challenge. In most areas with a Mediterranean-type climate, water is scarce. Thus, it becomes necessary to weigh the economic and environmental costs and benefits of heavily irrigated landscapes[144] and to consider recycled water systems.

For this technique, it might be helpful to explore alternative crops or cultivars that may not be traditionally grown in the area. For instance, low-flammability crops like bananas are being considered for Mediterranean climates. Crops like bananas have a high water content and are a high-profit plant. Furthermore, they can be planted on steep terrain and do not require more labor or machinery when compared to the maintenance of traditionally grown crops in the region. Additionally, cultivars of specific crops could be selected for the plant moisture content.[145]

Beyond crop selection, it is also important to consider understory maintenance. For instance, if unmaintained, the areas under plantations and vineyards could become overgrown with fast-growing, dry annual grasses. These plants ignite easily and quickly, and could significantly reduce the efficacy of the buffer.[146]

Lastly, agricultural stripping is just one type of greenbelt that could be used for wildfire risk reduction. Beyond food production, greenbelts can function as open space, parks, preserves, bike paths, playing fields, or golf courses.

# Notes

1   "Past Fire Activity," National Park Service, accessed November 1, 2022, https://www.nps.gov/yose/learn/nature/firehistory.htm

2   *Confronting the Wildfire Crisis: A Strategy for Protecting Communities and Improving Resilience in America's Forests* (USDA Forest Service, 2022).

3   Stephen Pyne, "Pyrocene Park," *AEON*, March 24, 2022, https://aeon.co/essays/what-yosemites-fire-history-says-about-life-in-the-pyrocene

4   Matt Plucinski, "Fighting Flames and Forging FIrelines: Wildfire Suppression Effectiveness at the Fire Edge," *Current Forestry Reports* 5 (January 2019): 1–19.

5   *Confronting the Wildfire Crisis.*

6   Pyne, "Pyrocene Park."

7   "Past Fire Activity."

8   Jan van Wagtendonk and James Lutz, "Fire Regime Attributes of Wildland Fires in Yosemite National Park, USA," *Fire Ecology* 3, no.2 (2007): 34–52.

9   "Chapter 4: Strategy and Tactics," International Fire Service Training Association, accessed November 1, 2022, https://www.ifsta.org/sites/default/files/GC_Ch_4.pdf

10  "Chapter 4: Strategy and Tactics."

11  Plucinski, "Fighting Flames."

12  Francisco Moreira et al., "Wildfire Management in Mediterranean-Type Regions: Paradigm Change Needed," *Environmental Research Letters* 15 (January 2020): 1–6.

13  Plucinski, "Fighting Flames."

14  *Confronting the Wildfire Crisis.*

15  Plucinski, "Fighting Flames."

16  *Confronting the Wildfire Crisis.*

17  Moreira, "Wildfire management."

18  "The New Generation Fire Shelter," National Wildfire Coordinating Group, accessed November 1, 2022, https://www.nwcg.gov/sites/default/files/publications/pms411.pdf

19  "Next Generation Structure Wrap," Firezat, Inc., accessed November 1, 2022, https://www.firezat.com/firezattestdatawebsite1.pdf

20  Raisa Bruner, "Protecting Homes from Wildfire with Aluminum Foil? A Tested Technology Gains Steam," *Time Magazine*, September 30, 2021, https://time.com/6103084/aluminum-house-fire-protection/?xid=homepage

21  Fumiaki Takahashi, "Whole-House Fire Blanket Protection from Wildland-Urban Interface Fires," *Frontiers in Mechanical Engineering* 5, no.60 (October 2019): 1–22.

22  Alejandra Borunda, "Wildfires Threaten the World's Oldest Trees - But Prescribed Burns are Protecting Them," *National Geographic*, September 21, 2021, https://www.nationalgeographic.com/environment/article/wildfires-threaten-the-worlds-oldest-trees-but-prescribed-burns-are-protecting-them

23  Bill Chappell, "Here's Why Firefighters are Wrapping Sequoia Trees in Aluminum Blankets," *NPR*, September 20, 2021, https://www.npr.org/2021/09/20/1038972507/california-sequoia-trees-general-sherman-aluminum-blanket#:~:text=A%20Single%20Fire%20Killed%20Thousands,fire%20scars%2C%22%20Christy%20M

24  Takahashi, "Whole-House."

25  "Next Generation."

26  Takahashi, "Whole-House."

27  Chappell, "Here's Why."

28  Takahashi, "Whole-House."

29  Bruner, "Protecting Homes."

30  "Next Generation."

31  Ibid.

32  Takahashi, "Whole-House."

33  "Collection of Wrapped Structure Photos Courtesy of Structure Protection Wrap Crew, FMO's, Archeaologists, and Volunteers," Firezat, Inc., accessed November 1, 2022, https://www.firezat.com/firezat_wrap_gallery_2020.pdf

34  Bruner, "Protecting Homes."

35  Chappell, "Here's Why."

36  Stephen Quarles et al., *Home Survival in Wildfire-Prone Areas: Building Materials and Design Considerations* (University of California Agriculture and Natural Resources, 2010).

37  Ibid.

38  Ibid.

39  *Is Your Home Hardened to Survive a Wildfire Ember Storm?* (California Fire Safe Council, 2019).

40  Quarles et al., *Home Survival.*

41  Ji Lee, Fangjiao Ma, and Yue Li, "Understanding Homeowner Proactive Actions for Managing Wildfire Risks," *Natural Hazards* 114, no.2 (2022): 1525-1547.

42  *Is Your Home Hardened.*

43  *Is Your Home Hardened.*

44  Quarles et al., *Home Survival.*

45  *Home Builder's Guide to Construction in Wildfire Zones* (Federal Emergency Management Agency, 2008).

46  Lee, "Understanding Homeowner."

47  Quarles et al., *Home Survival*.

48  *Home Builder's Guide*.

49  Mark Abadi and Daniel Allen, "An Engineer Spent 15 Years Fireproofing His California Home. Here's Why His House was the Last One Standing After a Devastating Blaze Last Year," *Business Insider Weekly*, February 12, 2020, https://www.businessinsider.com/california-fire-fireproof-home-sonoma-county-kincade-2020–2

50  Ibid.

51  Stephen Quarles, *Building a Wildfire-Resistant Home: Codes and Costs* (Headwaters Economics, 2018).

52  Ibid.

53  Ibid.

54  Christina Restaino et al., *Wildfire Home Retrofit Guide* (University of Nevada, Reno Extension, 2020).

55  Lee, "Understanding Homeowner."

56  Raphaele Blachi, "Surviving Bushfire: The Role of Shelters and Sheltering Practices during the Black Saturday Bushfires," *Environmental Science and Policy* 81 (2018): 86–94.

57  Alexandrea Spring, "Bushfire-Proof Houses are Affordable and Look Good - So Why aren't We Building More?" *The Guardian*, February 8, 2016, https://www.theguardian.com/sustainable-business/2016/feb/09/bushfire-proof-houses-black-saturday-innovations

58  Mario Arruda, Rogerio Tenreiro, and Federico Branco, "Rethinking Dwellings Protection for Wildland Fire in Southern Europe," *Journal of Performance of Constructed Facilities* 35, no.6 (2021): 1–17.

59  Ibid.

60  Kelly Hart, "Australian Fires Spark Interest in Underground and Mud Homes," accessed November 1, 2022, https://naturalbuildingblog.com/australian-fires-spark-interest-in-underground-and-mud-homes/#:~:text=The%20on%2Dgoing%20fire%20storm,since%20the%20bushfire%20crisis%20escalated

61  Joanne Brookfield, "Home in the Ground is Perfectly Sound," accessed November 1, 2022, https://www.domain.com.au/news/home-in-the-ground-is-perfectly-sound-20140222-33920/

62  "Bush Fire Resistant Houses," accessed November 1, 2022, https://www.baldwinobryan.com/bush-fire-resistant-houses.html

63  "House at Whittlesea VIC," accessed November 1, 2022, https://www.baldwinobryan.com/whittlesea-vic.html

64  Albert Fu, "The Facade of Safety in California's Shelter-in-Place Homes: History, Wildfire, and Social Consequence," *Critical Sociology* 39, no.6 (2012): 833–849.

65  Hart, "Australian Fires."

66  "Bush Fire Resistant."

67  Brookfield, "Home in the Ground."

68  Sarah Buckley, "Baldwin O'Bryan Beats Bushfires," accessed November 1, 2022, https://www.architectureanddesign.com.au/news/baldwin-o-bryan-bushfires

69  "Bush Fire Resistant."

70  Bushfire Prep.

71  Arruda et al., "Rethinking How."

72  Ibid.

73  Fu, "The Facade."

74  Brookfield, "Home in the Ground."

75  Stuart Rintoul, "Bunker Builders are Spreading like Wildfire," accessed November 1, 2022, https://www.wildfiresafetybunkers.com.au/pdf/Bunker_builders_spreading.pdf

76  Pierre Combrinck et al., "Challenges of Managing Fires along an Urban-Wildland Interface – Lessons from the Cape Peninsula, South Africa," accessed November 1, 2022, https://gfmc.online/wp-content/uploads/3-IWFC-077-Fowkes.pdf

77  *Cost-effectiveness of different fuel management measures in the Wildland-urban Interface* (GEF Fynbosfire Project, 2016).

78  Ibid.

79  Combrinck et al., "Challenges of Managing."

80  *Cost-effectiveness*.

81  Ibid.

82  *Cost-effectiveness*.

83  Ibid.

84  Ibid.

85  Ibid.

86  Combrinck et al., "Challenges of Managing."

87  *Cost-effectiveness*.

88  Ibid.

89  Ibid.

90  Robert Rowles, "Managing Hazards: Fire Management in the Cape Peninsula" (Msc dissertation, University of Johannesburg, 2012).

91  Hanna Prinssen, "A Fire-Scape: A New Form of a Fire Resilient Landscape" (MLA thesis, Academy of Architecture Amsterdam, 2020).

92 Ibid.

93 Ibid.

94 "Plan de Proteccion Contra Incendios Forestales Comuna Los Ángeles," accessed November 1, 2022, https://www.conaf.cl/wp-content/files_mf/1484084279PLANDEPROTECCIONCONTRAINCENDIOS FORESTALESLOSANGELES.pdf

95 Sandra Uribe, Christian Estades, and Volker Radeloff, "Pine Plantations and Five Decades of Land use Change in Central Chile," *PLoSOne* 15, no.3 (2020): 1-16.

96 Xavier Ubeda and Pablo Sarricolea, "Wildfires in Chile: A Review," *Global and Planetary Change* 146, (2016): 152–161.

97 Veronica Loewe, Victor Vargas, Juan Miguel Ruiz, Andrea Alvarez, and Felipe Lobo, "Creation and Implementation of a Certification System for Insurability and Fire Risk Classification for Forest Plantations," *USDA Forest Service Proceedings* (2015): 141–149.

98 Ubeda and Sarricolea, "Wildfires in Chile."

99 "Prevención de Incendios Forestales en Zonas de Interfaz de la Región del Biobío," accessed November 1, 2022, https://www.conaf.cl/incendios-forestales/prevencion/yo-tambien-soy-forestin-campana-de-prevencion-de-incendios-forestales-2020/prevencion-de-incendios-forestales-en-zonas-de-interfaz-de-la-region-del-biobio/

100 "Plan de Proteccion."

101 "Más de 20 km de cortafuegos construyó CONAF en zonas de interfaz de Los Ángeles," accessed November 1, 2022, https://www.prevencionincendiosforestales.cl/mas-de-20-km-de-cortafuegos-construyo-conaf-en-zonas-de-interfaz-de-los-angeles/

102 Ibid.

103 "Firewise Communities Recognition Program," Landworks, accessed November 17, 2021, http://landworksnpc.com/resource-centre/

104 Ibid.

105 Ibid.

106 "Technical Chart: National Fire Danger Index Rating System," Landworks, accessed November 17, 2021, http://landworksnpc.com/resource-centre/

107 Chandra Fick, *FireWise Communities Report 2016: FireWise Communities in the Fynbos Biome* (Cape Town: Kishugu PBO, 2017).

108 Ibid.

109 *A Guide to Integrated Fire.*

110 Steinberg, "Firewise."

111 Amanda Younge Hayes, *Lessons Learnt: FireWise Communities Workshop* (Worcester: Landworks, 2017): 1–27.

112 "Climate Change Adaptation in Kranshoek," Landworks, accessed November 17, 2021, https://landworksnpc.com/media/

113 Lucian Deaton, "Firewise in South Africa Making a World of Difference to Residents at Risk," National Fire Protection Association, accessed November 17, 2021, https://www.nfpa.org/News-and-Research/Publications-and-media/Blogs-Landing-Page/Fire-Break/Blog-Posts/2016/02/26/firewise-in-south-africa-making-a-world-of-difference-to-residents-at-risk

114 Fick, *FireWise.*

115 Steinberg, "Firewise."

116 Ibid.

117 Fick, *FireWise.*

118 Deaton, "Firewise."

119 *A Guide to Integrated Fire.*

120 Sarah Toth, "Pyro-Diversion: Planning for Fire in the San Gabriel Valley," accessed November 1, 2022, https://www.asla.org/2018studentawards/494988-Pyro_Diversion.html

121 Ibid.

122 Ibid.

123 Ibid.

124 Jon Keeley, Greg Rubin, Teressa Brennan, and Bernadette Piffard, "Protecting the Wildland-Urban Interface in California: Greenbelts vs Thinning for Wildfire Threats to Homes," *Bulletin, Southern California Academy of Sciences* 119, no.1 (2020): 35–47.

125 Ibid.

126 Ibid.

127 Ibid.

128 Ibid.

129 Ibid.

130 Esteve Canyameres et al., "Grinformed and Medifire, A project for the prevention of great forest fires in the Mediterranean" (Presentation, International Conference on Forest Fire Research, Coimbra, Portugal, 2006).

131 Andreu Palacios Megias, "Proposta de parcel·les de crema prescrita a l'Espai d'Interès Natural de la Serra de Llaberia," accessed November 1, 2022, https://repositori.udl.cat/handle/10459.1/49004

132 "Jornadas Tecnicas Selvicolas," accessed November 1, 2022, http://www.forestal.cat/bdds/imatges_db/biblioteca/BIBLIOTECA_DOCUMENT2_0359200013390721.pdf

133 Canyameres et al., "Grinformed."

134 "Jornadas Tecnicas Selvicolas."
135 Megias, "Proposta de parcel."
136 *The Critical Role of Greenbelts in Wildfire Resilience* (San Francisco: Greenbelt Alliance, 2021).
137 Ibid.
138 "Prevention Action Increases Large Fire Response Preparedness: Fuel Management Smart Solutions Towards Fire Resilient Landscapes," accessed November 1, 2022, https://www.prevailforestfires.eu/wp-content/uploads/2021/04/4.1.pdf
139 *The Critical Role.*
140 Xiao Fu, Abigail Lidar, Michael Kantar, and Barath Raghavan, "Edible Fire Buffers: Mitigation of Wildfire with Multifunctional Landscapes," accessed November 1, 2022, https://raghavan.usc.edu/papers/ediblefirebuffers-biorxiv21.pdf
141 Fu et al., "Edible Fire Buffers."
142 *The Critical Role.*
143 "Jornadas Tecnicas Selvicolas."
144 Claudia Herbert and Van Butsic, "Assessing the Effectiveness of Green Landscape Buffers to Reduce Fire Severity and Limit Fire Spread in California: Case Study of Golf Courses," *Fire* 5, no.44 (2022): 1-16.
145 Fu et al., "Edible Fire Buffers."
146 Ibid.

# Bibliography

Abadi, Mark and Daniel Allen, "An Engineer Spent 15 Years Fireproofing His California Home. Here's Why His House was the Last One Standing After A Devastating Blaze Last Year." *Business Insider Weekly*, February 12, 2020, https://www.businessinsider.com/california-fire-fireproof-home-sonoma-county-kincade-2020-2

Arruda, Mario, Rogerio Tenreiro, and Federico Branco, "Rethinking Dwellings Protection for Wildland Fire in Southern Europe." *Journal of Performance of Constructed Facilities* 35, no.6 (2021): 1–17.

Blachi, Raphaele, "Surviving Bushfire: The Role of Shelters and Sheltering Practices During the Black Saturday Bushfires." *Environmental Science and Policy* 81 (2018): 86–94.

Borunda, Alejandra, "Wildfires Threaten the World's Oldest Trees - But Prescribed Burns are Protecting Them." *National Geographic*, September 21, 2021, https://www.nationalgeographic.com/environment/article/wildfires-threaten-the-worlds-oldest-trees-but-prescribed-burns-are-protecting-them

Brookfield, Joanne, "Home in the Ground is Perfectly Sound." Accessed November 1, 2022. https://www.domain.com.au/news/home-in-the-ground-is-perfectly-sound-20140222-33920/

Bruner, Raisa, "Protecting Homes from Wildfire with Aluminum Foil? A Tested Technology Gains Steam." *Time Magazine*, September 30, 2021. https://time.com/6103084/aluminum-house-fire-protection/?xid=homepage

Buckley, Sarah, "Baldwin O'Bryan Beats Bushfires." Accessed November 1, 2022. https://www.architecturean-ddesign.com.au/news/baldwin-o-bryan-bushfires

"Bush Fire Resistant Houses." Accessed November 1, 2022. https://www.baldwinobryan.com/bush-fire-resistant-houses.html

Canyameres, Esteve, Francese Castro, Miguel Galante, Gael Rosello, Marco Marchetti, Nuria Ferrer, Susana Dias et al. "Grinformed and Medifire, A Project for the Prevention of Great Forest Fires in the Mediterranean." Presentation at the International Conference on Forest Fire Research, Coimbra, Portugal, 2006.

Chappell, Bill, "Here's Why Firefighters are Wrapping Sequoia Trees in Aluminum Blankets." *NPR*, September 20, 2021. https://www.npr.org/2021/09/20/1038972507/california-sequoia-trees-general-sherman-aluminum-blanket#:~:text=A%20Single%20Fire%20Killed%20Thousands,fire%20scars%2C%22%20Christy%20M

"Chapter 4: Strategy and Tactics." International Fire Service Training Association. Accessed November 1, 2022. https://www.ifsta.org/sites/default/files/GC_Ch_4.pdf

"Collection of Wrapped Structure Photos Courtesy of Structure Protection Wrap Crew, FMO's, Archaeologists, and Volunteers." Firezat, Inc. Accessed November 1, 2022. https://www.firezat.com/firezat_wrap_gallery_2020.pdf

Combrinck, Pierre, Yolande Dwarika, Sandra Fowkes, Philip Prins, and P. Smith, "Challenges of Managing Fires along an Urban-Wildland Interface – Lessons from the Cape Peninsula, South Africa." Accessed November 1, 2022. https://gfmc.online/wp-content/uploads/3-IWFC-077-Fowkes.pdf

*Confronting the Wildfire Crisis: A Strategy for Protecting Communities and Improving Resilience in America's Forests* (USDA Forest Service, 2022).

*Cost-Effectiveness of Different Fuel Management Measures in the Wildland-Urban Interface.* Cape Town: GEF Fynbosfire Project, 2016.

Deaton, Lucian, "Firewise in South Africa Making a World of Difference to Residents at Risk." Accessed November 17, 2021. https://www.nfpa.org/News-and-Research/Publications-and-media/Blogs-Landing-Page/Fire-Break/Blog-Posts/2016/02/26/firewise-in-south-africa-making-a-world-of-difference-to-residents-at-risk

Fick, Chandra, *FireWise Communities Report 2016: FireWise Communities in the Fynbos Biome.* Cape Town: Kishugu PRO, 2017.

Fu, Albert, "The Facade of Safety in California's Shelter-in-Place Homes: History, Wildfire, and Social Consequence." *Critical Sociology* 39, no.6 (2012): 833–849.

Fu, Xiao, Abigail Lidar, Michael Kantar, and Barath Raghavan, "Edible Fire Buffers: Mitigation of Wildfire with Multifunctional Landscapes." Accessed November 1, 2022. https://raghavan.usc.edu/papers/ediblefirebuffers-biorxiv21.pdf

Hart, Kelly, "Australian Fires Spark Interest in Underground and Mud Homes." Accessed November 1, 2022. https://naturalbuildingblog.com/australian-fires-spark-interest-in-underground-and-mudhomes/#:~:text=The%20on%2Dgoing%20fire%20storm,since%20the%20bushfire%20crisis%20escalated

Hayes, Amanda Younge, *Lessons Learnt: FireWise Communities Workshop.* Worcester: Landworks, 2017.

Herbert, Claudia and Van Butsic, "Assessing the Effectiveness of Green Landscape Buffers to Reduce Fire Severity and Limit Fire Spread in California: Case Study of Golf Courses." *Fire* 5, no. 44(2022): 1–16.

*Home Builder's Guide to Construction in Wildfire Zones* (Federal Emergency Management Agency, 2008).

"House at Whittlesea VIC." Accessed November 1, 2022. https://www.baldwinobryan.com/whittlesea-vic.html

*Is your Home Hardened to Survive a Wildfire Ember Storm?* (California Fire Safe Council, 2019).

"Jornadas Tecnicas Selvicolas." Accessed November 1, 2022. http://www.forestal.cat/bdds/imatges_db/biblioteca/BIBLIOTECA_DOCUMENT2_0359200013390721.pdf

Keeley, Jon, Greg Rubin, Teressa Brennan, and Bernadette Piffard, "Protecting the Wildland-Urban Interface in California: Greenbelts vs Thinning for Wildfire Threats to Homes." *Bulletin, Southern California Academy of Sciences* 119, no.1 (2020): 35–47.

Landworks, "Climate Change Adaptation in Kranshoek." Accessed November 17, 2021. https://landworksnpc.com/media/

Landworks, "Firewise Communities Recognition Program." Accessed November 17, 2021. http://landworksnpc.com/resource-centre/

Landworks, "Technical Chart: National Fire Danger Index Rating System." Accessed November 17, 2021. http://landworksnpc.com/resource-centre/

Lee, Ji, Fangjiao Ma, and Yue Li, "Understanding Homeowner Proactive Actions for Managing Wildfire Risks." *Natural Hazards* 114, no.2 (2022): 1525–1547.

Loewe, Veronica, Victor Vargas, Juan Miguel Ruiz, Andrea Alvarez, and Felipe Lobo, "Creation and Implementation of a Certification System for Insurability and Fire Risk Classification for Forest Plantations." *USDA Forest Service Proceedings* (2015), 141–149.

"Más de 20 km de cortafuegos construyó CONAF en zonas de interfaz de Los Ángeles." Accessed November 1, 2022. https://www.prevencionincendiosforestales.cl/mas-de-20-km-de-cortafuegos-construyo-conaf-en-zonas-de-interfaz-de-los-angeles/

Megias, Andreu Palacios, "Proposta de parcel·les de crema prescrita a l'Espai d'Interès Natural de la Serra de Llaberia." Accessed November 1, 2022.

Moreira, Francisco, Davide Ascoli, Hugh Safford, Mark Adams, Jose Moreno, Jose Pereira, Filipe Catry, et al., "Wildfire Management in Mediterranean-Type Regions: Paradigm Change Needed." *Environmental Research Letters* 15 (January 2020): 1–6.

"Next Generation Structure Wrap." Firezat, Inc. Accessed November 1, 2022. https://www.firezat.com/firezat-testdatawebsite1.pdf

"Past Fire Activity." National Park Service. Accessed November 1, 2022. https://www.nps.gov/yose/learn/nature/firehistory.htm

"Plan de Protección Contra Incendios Forestales Comuna Los Ángeles." Accessed November 1, 2022. https://www.conaf.cl/wp-content/files_mf/1484084279PLANDEPROTECCIONCONTRAINCENDIOSFORESTALESLOSANGELES.pdf

Plucinski, Matt, "Fighting Flames and Forging Firelines: Wildfire Suppression Effectiveness at the Fire Edge." *Current Forestry Reports* 5 (January 2019): 1–19.

"Prevention Action Increases Large Fire Response Preparedness: Fuel Management Smart Solutions Towards Fire Resilient Landscapes." Accessed November 1, 2022. https://www.prevailforestfires.eu/wp-content/uploads/2021/04/4.1.pdf

"Prevención de Incendios Forestales en Zonas de Interfaz de la Región del Biobío." Accessed November 1, 2022. https://www.conaf.cl/incendios-forestales/prevencion/yo-tambien-soy-forestin-campana-de-prevencion-de-incendios-forestales-2020/prevencion-de-incendios-forestales-en-zonas-de-interfaz-de-la-region-del-biobio/

Prinssen, Hanna, "A Fire-Scape: A New Form of a Fire Resilient Landscape." MLA thesis, Academy of Architecture Amsterdam, 2020.

Pyne, Stephen, "Pyrocene Park." *AEON*, March 24, 2022. https://aeon.co/essays/what-yosemites-fire-history-says-about-life-in-the-pyrocene

Quarles, Stephen, *Home Survival in Wildfire-Prone Areas: Building Materials and Design Considerations* (University of California Agriculture and Natural Resources, 2010).

Quarles, Stephen, Yana Valachovic, Gary Nakamura, Glenn Nader, and Michael de Lasaux, *Building a Wildfire-Resistant Home: Codes and Costs* (Headwaters Economics, 2018).

Restaino, Christina, Susan Kocher, Nicole Shaw, Steven Hawks, Carlie Murphy, and Stephen Quarles, *Wildfire Home Retrofit Guide* (University of Nevada, Reno Extension, 2020).

Rintoul, Stuart, "Bunker Builders are Spreading like Wildfire." Accessed November 1, 2022. https://www.wildfiresafetybunkers.com.au/pdf/Bunker_builders_spreading.pdf

Rowles, Robert, "Managing Hazards: Fire Management in the Cape Peninsula." Msc dissertation, University of Johannesburg, 2012.

Spring, Alexandrea, "Bushfire-Proof Houses are Affordable and Look Good - So Why aren't We Building More?" *The Guardian*, February 8, 2016. https://www.theguardian.com/sustainable-business/2016/feb/09/bushfire-proof-houses-black-saturday-innovations

Takahashi, Fumiaki, "Whole-House Fire Blanket Protection from Wildland-Urban Interface Fires." *Frontiers in Mechanical Engineering* 5, no.60 (October 2019): 1–22.

*The Critical Role of Greenbelts in Wildfire Resilience* (Greenbelt Alliance, 2021).

"The New Generation Fire Shelter." National Wildfire Coordinating Group. Accessed November 1, 2022. https://www.nwcg.gov/sites/default/files/publications/pms411.pdf

Toth, Sarah, "Pyro-Diversion: Planning for Fire in the San Gabriel Valley." Accessed November 1, 2022. https://www.asla.org/2018studentawards/494988-Pyro_Diversion.html

Ubeda, Xavier and Pablo Sarricolea, "Wildfires in Chile: A Review." *Global and Planetary Change* 146 (2016): 152–161.

Uribe, Sandra, Christian Estades, and Volker Radeloff, "Pine Plantations and Five Decades of Land Use Change in Central Chile." *PLoS One* 15, no.3 (2020): 1–16.

Wagtendonk, Jan van and James Lutz, "Fire Regime Attributes of Wildland Fires in Yosemite National Park, USA." *Fire Ecology* 3, no.2 (2007): 34–52.

# Co-Creation

Figure 5.1
View of a landscape and structural remnants around Lake Berryessa after the LNU Lightning Complex Fires of 2020.

Figure 5.2
A cultural burn conducted at Cache Creek Preserve in the fall of 2022.

Figure 5.3
A powerline running through the Caribou-Targhee National Forest.
Photograph by US Forest Service.

Figure 5.4
A wetland in Cache Creek Preserve following a cultural burn to encourage resprouting.

Figure 5.5
A feller buncher thins trees to reduce the threat of wildfire near a community.
Photograph by Neal Herbert, National Park Service.

Figure 5.6
A flock of sheep used for prescribed grazing on BLM-managed public lands near the South Fork American River.
Photograph by Monte Kawahara, Bureau of Land Management.

# Chapter 5
# Co-Creation

*to create (something) by working with one or more others; to create (something) jointly; engaging in an intentional relationship in order to make something together.*

These are approaches to fire and fire stewardship that embrace and utilize landscape forces, while also trying to intentionally guide them. A co-creative approach is one of give and take and feedback between people and landscapes. It's an approach that understands that landscapes cannot be controlled, but can be stewarded, cared for, and collaborated with. Co-creation implies a lack of clear and distinct authorship of these techniques, as agency is broadly shared and distributed across landscape assemblies, people, and climates.

DOI: 10.4324/9781003172956-7

1 mi

Adelaide

*Patchy Planting*

Mount Lofty Ranges

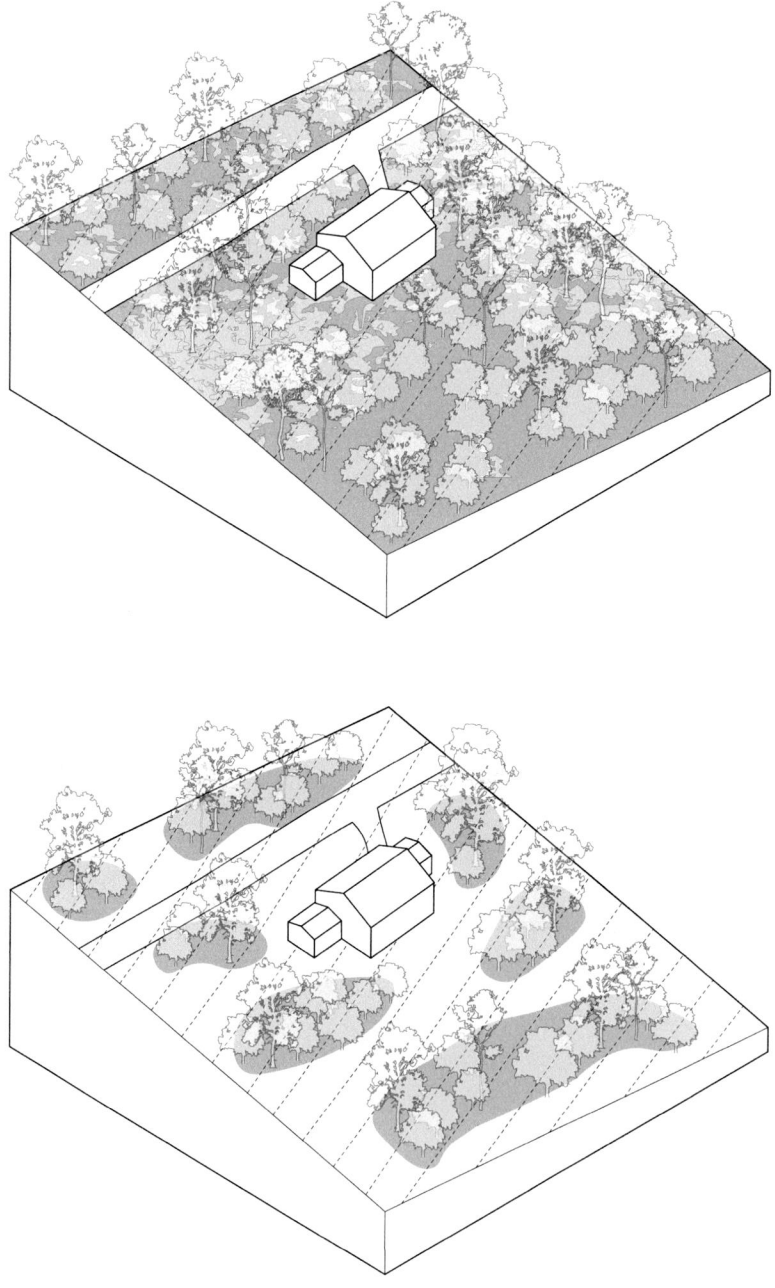

# 12

*Patchy Planting*

| | |
|---|---|
| Description | Modifying the spatial arrangement of vegetation around structures |
| Example Location | Adelaide, Australia |
| Size | Variable |
| Implementer | Fenner School of Environment and Society at the Australian National University |
| Team | University of California Cooperative Extension Division of Agriculture and Natural Resources, Bren School of Environmental Science & Management, Earth Research Institute UC Santa Barbara |

## Technique Overview

In Adelaide, Australia, wildfire-related structure losses are increasing due to growth in the wildland-urban interface (WUI). There are three primary ways in which ignition can happen. The first and most prevalent way, is through ember attacks when small burning pieces of wood or plant material are blown ahead of the fire front and create spot fires. These embers easily get caught in architectural and outdoor features. The second way is through surface fires which generally travel near to the ground if there is uninterrupted vegetation around the home. The third way is through intense radiation either from crown fires, or from nearby structure or car fires.

In an attempt to reduce the risk of structural loss, it is often advised to simply remove vegetation close to homes. The regulations that support this recommendation are often generic, focusing solely on the clearing of vegetation to set distances from structures.[1]

In 2017, a group of researchers sought to challenge this widely accepted practice by asking how those living in the WUI might better balance wildfire safety with the many benefits of having vegetation nearby the home. These benefits include increased privacy, aesthetics, biodiversity, and reduced energy consumption.[2]

Figure 5.7 (*Previous*) Aerial of Adelaide. Source: Google Earth 2022.

Figure 5.8 (*Left*) A view of a home with continuous vegetation all around it (top) and the same view with discrete patches of vegetation to reduce horizontal and vertical fire spread (bottom).

## Planning and Design Process

To help answer this question, the researchers studied 499 residences impacted by three wildfires that ignited in Southeastern Australia in February 2009 – the East Kilmore, the Murrindindi, and the Churchill fires. In total, these wildfires burned over 480,000 acres of forested landscape (consisting primarily of pine and eucalyptus plantations), rural, semi-rural, and urban land and 1,925 houses. Using aerial imagery and geographic information systems (GIS) the team documented the damage of the fires on the structures approximately two weeks following the burn.[3]

Three primary hypotheses drove their study. First, they hypothesized that homes with many discrete patches of vegetation around structures fared better than those with fewer patches and larger swaths of continuous vegetation. The second hypothesis was that homes with greener vegetation (vegetation with high moisture content) fared better than those without. The third hypothesis was that homes with less vegetation upwind of prevailing winds during fire season fared better than homes with more.[4]

## Performance and Evaluation

The data gathered from the study supported their three hypotheses. First, they found that the spatial arrangement of vegetation 140 feet out from a home related to the likelihood it would survive a fire. Houses with many patches or islands of vegetation fared better than those that did not. They found that these patches created horizontal breaks which helped to reduce the intensity and spread of a fire. Homes with continuous, large blocks of vegetation, on the other hand, were more vulnerable.

They also found that homes with greener vegetation fared better than those that did not. These greener plants generally had a high moisture content and had lower levels of flammability, either due to plant physiology or through supplemental irrigation provided by the homeowner.

Lastly, they found that homes with less vegetation upwind of the dominant wildfire direction were at less risk than those with more vegetation. The reason behind this is that fire tends to travel faster downwind because vegetation (fuel) is preheated and embers tend to travel in this direction.[5]

These findings give homeowners in the WUI more options for implementing wildfire risk reduction measures on their properties. Allowing for some vegetation to exist around homes can improve aesthetics, create more privacy, support biodiversity, reduce energy consumption, increase property values, and reduce risk reduction costs. Perhaps new policies based on these findings might encourage more people who are currently reluctant to remove vegetation around their homes, to implement some risk reduction measures?[6]

## Challenges and Future Research

While these findings are promising for residents unwilling to drastically remove vegetation around their homes, there are additional aspects to consider.

When it comes to vegetation patchiness, one should also consider how larger clusters of vegetation might actually create a sheltering effect during fires. Additionally, it is important to think about what happens between the patches, as combustible material should be avoided. Also, patchiness is important not only for creating horizontal breaks but also for creating vertical breaks. It is necessary to reduce ladder fuels as much as possible. Lastly, seasonal pruning is an important (yet often, overlooked) exercise for maintaining yard patchiness.[7]

When it comes to increasing the greenness of plants around homes, it should be noted that most drought-tolerant native plants often do not fit within this category. Also, greenness may also relate to the amount of irrigation a yard receives, which may pose problems in areas with low water availability. Lastly, ensuring plants remain green over time often requires routine maintenance efforts which can be time- and resource-intensive.[8]

For the third finding regarding upwind vegetation, it should be noted that downwind vegetation can also pose a risk to homes as wildfire movement can be unpredictable. Also, if there is a slope in the upwind direction, the recommended spacing between plants and trees should increase with the grade change.[9]

Changing the spatial arrangement of vegetation around homes, while important, is not the only factor in reducing structural loss. For example, hardening homes is another key consideration. And lastly, this technique loses efficacy as the severity of wildfire events increases.[10]

1 mi

Anglesea Heath

Torquay

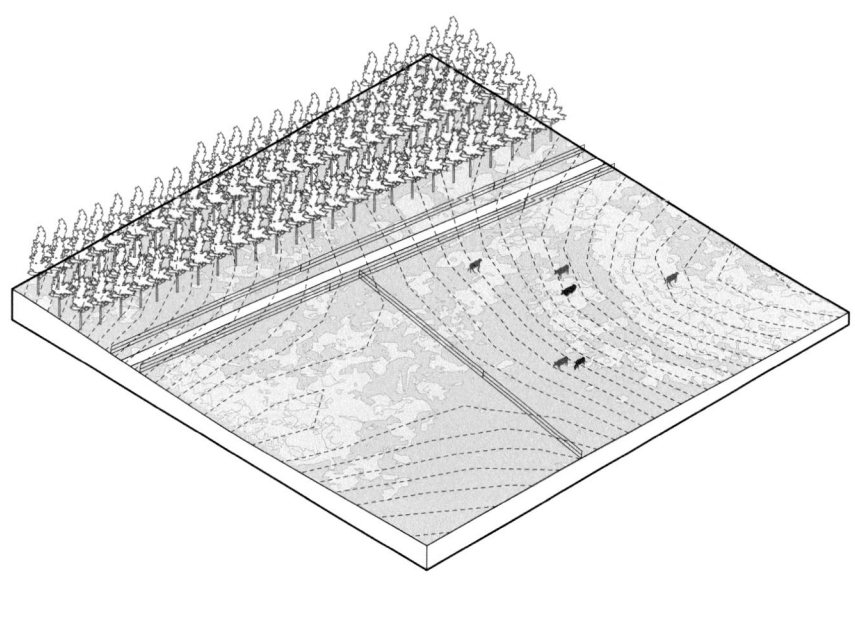

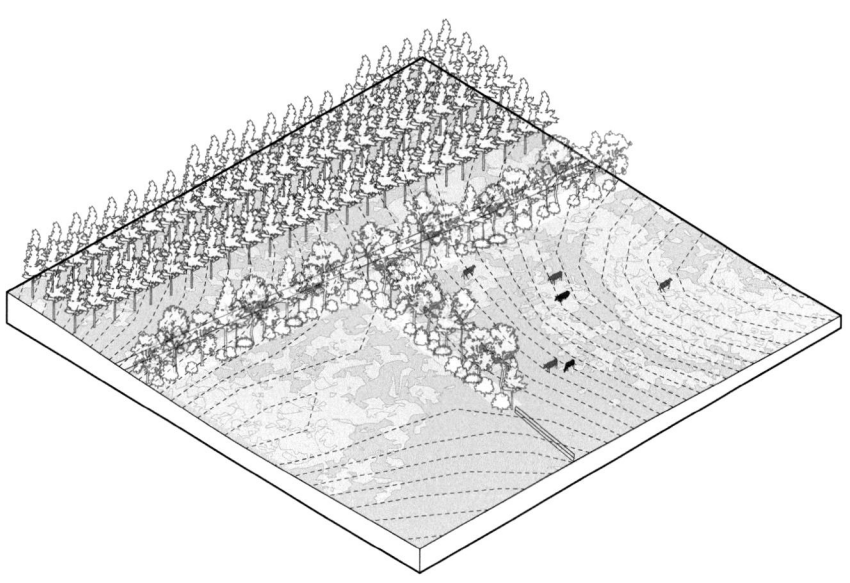

# 13

## Ember Trapping

| | |
|---|---|
| Description | Planting rows of vegetation to slow wind, absorb heat, and catch embers |
| Example Location | Victoria, Australia |
| Size | Variable |
| Primary Implementer | Turner Cattle Ranch |
| Stakeholders and Team Members | Landcare groups, local native plant nurseries, neighbors |

## Technique Overview

In 2017, Dean Turner, an outdoor education and sustainable design practitioner, began managing his family cattle farm outside of Melbourne. Turner, alongside his brothers, decided to plant over 10,000 trees in shelterbelts on their farm. In a general sense, shelterbelts are strips of living vegetation – either planted or retained – that can be used for a range of functions including herbivore protection and timber production.[11] They are also designed to serve as a windbreak, creating a calm, protected area on the leeward side of the strip.[12]

If designed carefully, shelterbelts also have the potential to reduce the impact of bushfires. First, they can slow the speed of an approaching fire in the case that it spreads to the leeward side. If the velocity is reduced by 50%, then the fire will be moving at half of its pre-shelterbelt speed.[13] Shelterbelts can also act as a physical barrier for direct flame contact and radiant heat emanating from a fire. This is especially helpful if the property has sensitive buildings or infrastructure.[14] Lastly, they can catch and extinguish embers carried ahead of or behind a fire, as embers tend to drop due to a decrease in wind velocity.[15] In this case, it is helpful for the belts to have fire-resistant foliage, non-flammable leaf litter, and high soil moisture content.[16]

## Planning and Design Process

When planning their shelterbelt system, the Turner family weighed a range of design considerations. The first consideration was the height of the shelterbelt which corresponds to the size of the protected area on the leeward side, typically 10–15 times the average height of the vegetation.[17] To achieve this protection quickly, the family chose taller and faster-growing species for part of the belt.[18] They also selected hardy species with a high survival rate, those that provided habitat value for local fauna, and those with fire-resistant qualities. Another consideration was length and form. Typically, shelterbelts shorter than 10 times the mature height of their vegetation can actually accelerate wind laterally and create turbulence at the end of the strips. Thus, the family developed longer shelterbelts and oriented them in a T-shaped form in case wind patterns unexpectedly shift during a bushfire event.[19] Shelterbelt continuity is another consideration, as gaps in the system can reduce efficacy. To ensure uniformity, the Turner family created a shelterbelt system that contained three to five rows of vegetation, with a particular focus on branching and foliage uniformity.[20] They positioned taller trees in the interior rows and smaller shrubs in the exterior rows, both with a spacing of five to ten feet on-center.[21] They also aimed for a consistent vegetative density of 50–60% from the ground plane to the tops of the trees. By promoting semi-permeability, they allowed for some air to pass through the system, while still maintaining a significant protective zone.[22]

After removing weeds and prepping the soil to encourage water penetration, the family erected a fence about five feet from the shelterbelt to prevent herbivores and wildlife from damaging the young plants. To do this, they used wire netting and solid fencing material and provided access gates to allow for future grazing once the plants are established. Then, they took seedlings grown in individual pots known as tubestock and planted them in an alternating fashion.[23]

The shelterbelts planted on Dean Turner's family farm in Victoria are relatively young; thus, the family is waiting for them to grow in order to assess their efficacy and to determine future management considerations and the possibility for expansion. In the meantime, Turner and his brothers are contemplating ways in which they can keep the shelterbelts hydrated so that the plants will be less likely to ignite. Dean Turner is also developing a series of workshops aimed at designing and planting fire-retardant shelterbelts. This effort is being spearheaded by Turner's environmental organization, The Crossing Land, located near Bermagui, Australia on the coast of Southern New South Wales.

## Performance and Evaluation

Beyond decreasing wind speeds and slowing the progress of an impending fire, shielding elements from radiant heat and direct flames, and trapping firebrands, shelterbelts can provide a range of other benefits. First, they can increase crop production by reducing onsite moisture loss and the potential for erosion. This is especially important during the summer months, when hot, dry winds can negatively impact productive land.[24]

Shelterbelts can also support herbivores by creating protective spaces during extreme weather events. They can also protect younger animals that tend to be more vulnerable. Furthermore, shelterbelts can be designed as an additional or emergency food source for herbivores, either through grazing or harvesting.[25]

Augmenting habitat is another benefit associated with shelterbelts, serving as a protective space, food source, and migration corridor for a range of species. Things to consider when planting a shelterbelt for habitat creation include: using native grasses, shrubs and trees, establishing a wide belt (at least five to seven rows), and linking the belt to existing patches or corridors.[26] On a related note, shelterbelts can also promote natural pest control by attracting wildlife, like birds and bats, that consume insect pests or carry diseases that are detrimental to insect pests. This is considered a welcome alternative to conventional pesticides, which can be expensive and can negatively impact the landscape.[27]

Additionally, the creation of shelterbelts can increase the value of a property by improving landscape aesthetics and bolstering the local identity of the site.[28]

Lastly, shelterbelts can also be used to produce a number of goods including fruits, nuts, flowers, and high-value timber. High-value timber is usually possible in shelterbelts with five or more rows and is typically grown with other plant material. This kind of timber can be used for firewood or fences.[29]

## Challenges and Future Research

While carefully designed shelterbelts can be successful at reducing the risk of damage from bushfires and have a range of other benefits, there are many challenges associated with implementation. First, it is important to clarify that no plant or tree is truly fireproof. While some are difficult to ignite, they can still burn under extreme weather conditions.[30]

Weeds are another challenge when designing and planting shelterbelts, as they can rapidly colonize an area, reduce habitat value, and prevent the regeneration of shelterbelt vegetation. To mitigate the effects of weeds on the shelterbelt system, it is helpful to mechanically remove weeds prior to flowering to prevent further spread. Furthermore, when deciding the location of the shelterbelts, it is helpful to avoid places treated with fertilizer, as those zones can be prone to invasive alien species.[31]

The maintenance of shelterbelts is critical for successfully reducing risk. First, it is important to inspect trees and remove any loose or flaky bark. It is also helpful to prune trees so that there are no branches touching the understory. If any trees form cavities, the cavities should be filled. Furthermore, all dead trees or branches should be removed from the belt. Additionally, any grasses that grow in the belt should be regularly cut or slashed. Lastly, any accumulated leaf litter should be removed. All of this debris can be composted into the ground to improve soil health.[32]

If not planted with enough species, shelterbelts can also be susceptible to pests and disease. These attacks can cause die-off and create undesired fuel in the shelterbelt as well as gaps in the system.[33]

In order to increase the survival rates of shelterbelt plants, it is helpful to keep them well-hydrated. This is especially useful following the initial planting period when species are particularly vulnerable to desiccation. During this time, hand watering is common – a task which can be time- and resource-intensive. Even after plants are established, supplemental watering can be helpful to ensure survivability and reduce flammability. To support this effort, some form of irrigation must be available for use.[34]

Planning for access through shelterbelts is yet another challenge. Since large gaps within shelterbelts reduce efficacy, it can be difficult to plan for vehicular or even pedestrian access. As a result, these points of access are typically located at the ends of belts, which complicates site circulation needs.[35]

Shelterbelts, especially wide ones, can also take up a significant amount of productive farmland, which can be a deterrent for many property owners. Furthermore, if a property is large, multiple belts may be required to provide sufficient protection.[36]

Lastly, while there are significant data on the efficacy of shelterbelts in reducing wind speeds and providing agricultural protection, there are not a lot of data related to their role in reducing risk during bushfire events. In Victoria and New South Wales, much of the information on this topic is anecdotal and based on observations.[37]

*Ember Trapping Interview*

Figure 5.11 Dean Turner, Outdoor Educator and Sustainable Design Practitioner.

DESIGN BY FIRE: Can you tell me a bit about yourself and about the mission of your organization?

DEAN TURNER: I run an outdoor education, environmental education, and sustainable design center called The Crossing Land, for the local community and especially for young people. I've been doing that for 22 years now, and before that I had a background in agricultural science and came off the land from family farms. One of the biggest learnings for me along the way has been the incorporation of permaculture. And the mission of The Crossing Land is really about sustainable design leadership and land care.

DF: Can you provide some details about the shelterbelt you implemented at your family's farm in Victoria? What was implemented there and what the design considerations were for that project?

DT: That particular shelterbelt is about 3 years old. At The Crossing we've put in some as well, and at my home, and those are about a year old. We've had the best season ever as far as rainfall, so everything that we've put in this year should survive just fine. As we go along, we plan to assess the wind amelioration of each belt at various distances over time on strong wind days. Additionally, we'd want to assess moisture retention at various distances during hot periods in order to estimate the retardant capabilities of the shelter belts.

There has been a lot of research on what were originally called windbreaks, which I'm choosing to call shelterbelts now. There are a number of studies on them showing notable pressure differences on either side of the windbreak. So, we know we can design them to be semi-permeable to achieve maximum benefits as far as wind pressure, and now we just need the next step as far as how we can reduce flammability. I used to think it was all about species selection, but I'm now changing my mind a bit about that. I think species selection is important, and it can help, but the really important thing I think is that everything is well-watered. There should be a supplemental water system in place. I think equally important is the use of indigenous, native species. And there's also the point of soil moisture conservation being a vitally important part of design for arresting wildfires and diverting them.

DF: Could you talk a bit more about some of the benefits of working with shelterbelts that we haven't touched on?

DT: There's a long list of benefits that I can touch on including carbon capture, creating drought reserve paddocks, generating additional income through managed planting and harvesting of high-value timber within the belt, increasing natural pest control which adds fertilizer to ground from droppings, and creating habitat for bees which help to pollinate the landscape.

DF: Are there any other challenges that would prevent people from building them?

DT: The main things preventing people are probably space, cost, fencing, and maintenance. Unfortunately, you do have to keep the space well-maintained for it to be fully beneficial. And then there's the logistics to keep the system watered at full capacity. My current project involves bringing water across the landscape to the shelterbelt using swales. It's also been a very slow uptake for large scale examples that we can look to. We're getting increasing bushfire recovery government support for the workshops, which is great, but we need more capacity for the prevention of fires through strategies like these as well.

DF: What are some future goals of this project and where do you see room for growth?

DT: This project is relatively unknown since we're working very locally at this point. Until our current shelterbelt examples become more mature, it will be difficult to share their importance and function at a wider scale. In the future we'd like to see the establishment of a series of shelterbelts in the district on larger properties, and in public areas such as around a showground. We also want to see the concepts of fire retardant landscapes applied more in general, especially in and around human settlements. I think the work we're doing has the potential to be scaled up and replicated in other places where wildfire is an issue.

DF: Do you think your work has contributed to community awareness about ecology, landscape maintenance, and the underlying issue of wildfire?

DT: It is beginning to. Our workshops on cool burning, fire preparation, and fire retardant landscapes are improving with each delivery and what began as three separate half day workshops is now being integrated into one weekend workshop. I think a leadership program such as the one we will pilot in 2022 will be a great way to spread the word and inspire further pilot sites and designs. I believe it will also help attract more media and academic attention.

1 mi

Drôme River

Allex

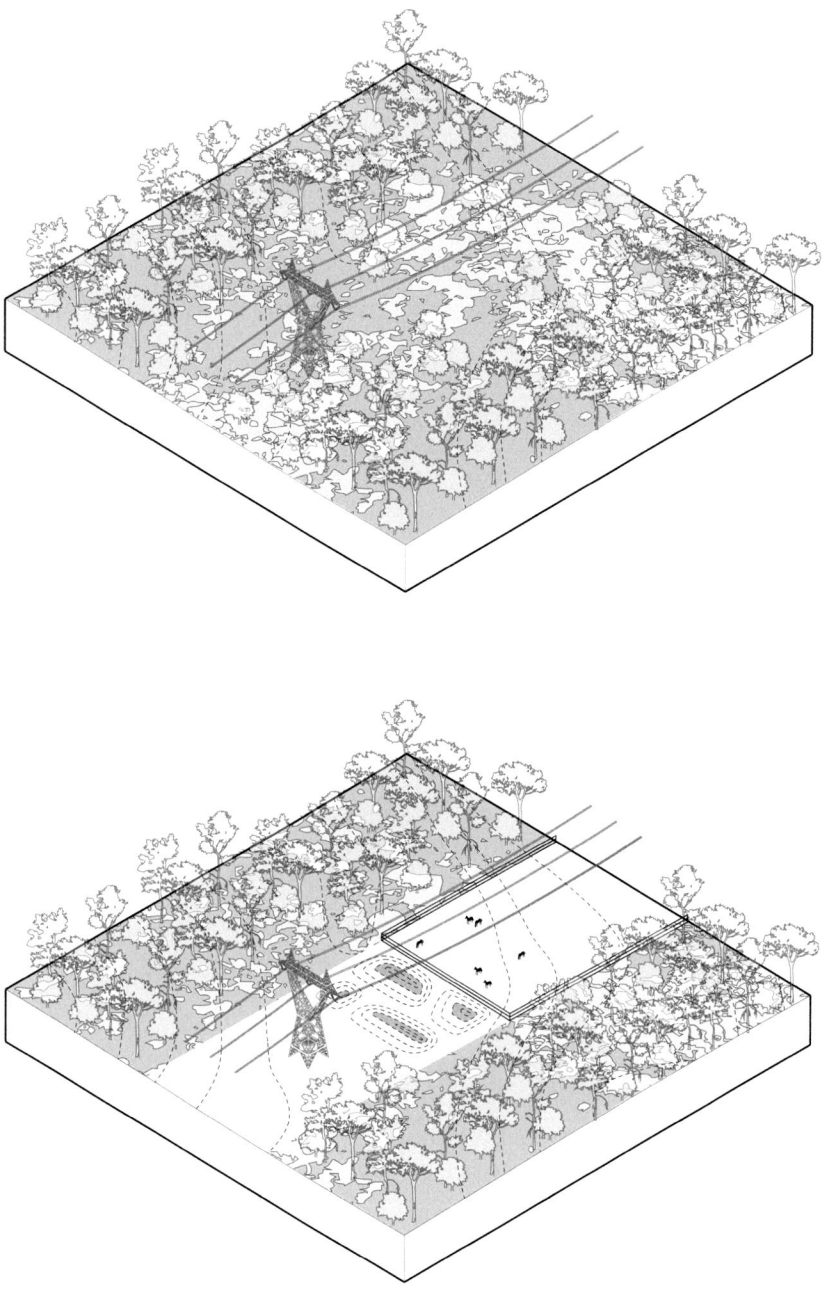

14

*Infrastructure Shadowing*

| | |
|---|---|
| Description | Creating wetland habitat and a grazing program under high voltage power lines |
| Example Location | Drôme, France (and 34 other sites in France and Belgium) |
| Size | 6 acres (pilot), 1,300 acres (current estimate) |
| Implementer | LIFE Elia – RTE |
| Team | Réseau de Transport d'Electricité (RTE), public and private landowners, transmission system operators (TSOs), local communities of Val de Drôme, Grane and Allex, farming operators, hunting and forestry organizations, and conservation groups |

## Technique Overview

In areas directly below and adjacent to high-voltage (HV) electricity lines, dead or dying trees can increase wildfire risk.[38] Additionally, issues such as line swaying due to wind and line lengthening due to heat can further increase risk.

To reduce risk, it is common for transmission system operators (TSOs) in Europe to develop linear safety corridors below HV power lines – sometimes over 150 feet in width – to deal with problematic vegetation.[39] Their approach is to target young vegetation every 3–12 years with a rotary cutter-equipped tractor. If a site is too steep or has too many physical obstacles, TSOs work manually. In both situations, mulched trees and shrubs are left onsite, typically scattered across the ground.[40] While this clear-cutting and mulching process succeeds in immediately removing <u>fuel</u> from the safety corridor, it poses challenges for long-term management. With access to light and soil enriched by mulched vegetation, seeds of pioneer species are able to quickly germinate and grow.[41]

In an effort to find alternatives to this conventional <u>fuel reduction</u> process, the LIFE Elia-RTE project emerged in 2011 and focused on 35 HV power line sites spread across both Belgium and France.[42] The project had three main goals: (1) create linear corridors along HV power lines to promote habitat creation and to bolster biodiversity, (2) experiment with new fuel reduction techniques for wildfire risk reduction, and (3) educate and raise public awareness about alternative management for electrical infrastructure.[43]

Figure 5.12 (*Previous*) Aerial of Drôme. Source: Google Earth 2022.

Figure 5.13 (*Left*) A view of an HV powerline corridor with significant problematic vegetation (top) and the same view with fuel reduction efforts, habitat creation, and grazing space (bottom).

## Planning and Design Process

This case study focuses on one HV power line site located just north of the Drôme River and south of the small community of Allex, France. When the land was purchased, it was overgrown with woody species and <u>invasive alien species</u> like *solidago goldenrod*.[44]

Once purchased, the team used heavy machinery to remove large woody plants and to prepare the landscape. The north side of the site was prepared for horse grazing and the south side of the site for wetland habitat creation. To create the wetland habitat, the team first looked for areas that were naturally wet or marshy or areas where rainwater run-off could help supply water to the ponds. In these areas, the soil was assessed for its ability to retain water. The team also focused on preserving easy access for maintenance and monitoring vehicles. Thus, they looked for areas more than 65 feet away from electrical pylons to create circulation buffers. They also considered access to sunlight, ways to reduce natural sedimentation, and how to avoid disrupting rare or endangered species.[45]

In general, most of the ponds designed for the project were around 1,000 to 1,500 square feet in size. If the terrain was flat, the team created depressions oriented in an east-west direction. If the terrain was sloped, the team positioned the pond so that the longer axis ran perpendicular to the slope. All ponds were designed to be curvilinear with irregular contours, often in a kidney bean-like shape. These design considerations supported the creation of microhabitat heterogeneity.[46] All slopes were designed to be no more than 5% in grade in order to allow for the establishment of plants, easy faunal access, and naturally warmed areas. Additionally, the lowest point in each pond tended to be about four feet below grade to create a refuge in times of extreme drought.[47]

Once sited and designed, the backhoe excavation work was carried out on dry or frozen ground to avoid overly disturbing the worksite. Then the team flattened the excavated soil under the power lines and let the depressions develop and evolve on their own. Over time, the depressions filled with water, plants colonized the edges, and fish, amphibians, insects, and other fauna emerged.[48]

After the ponds were fully excavated, local farmers were able to graze their animals in the linear corridor under the power lines. The north side was broken into two main grazing pastures. To contain the herbivores, temporary electrical fences were erected around the pastures and a small shelter was built to provide protection. The electrical fences prevented the herbivores from interfering with the wetland habitat to the south.

During the first grazing season, a number of horses owned by a local breeder were brought to the site to reduce the amount of woody vegetation within the safety corridor of HV power lines. Following the first season, the number of horses grazing the site was reduced to two. This grazing pattern repeated four times until 2017, which marked the end of project funding. Today, the site continues to be annually mowed and monitored.

By the end of the LIFE Elia-RTE project in 2017, over 1,300 acres of land under 85 miles of power lines in Belgium and France had been managed using unconventional procedures like the one outlined in this case study. Of these 1,300 acres, approximately six acres had been transformed into pond habitat in Drôme. Elsewhere, 675 acres had been restored as forest edges; 60 acres had been turned into orchards; 250 acres had been turned into bogs, grasslands, and meadows; 170 acres had been used for pasture; 70 acres had combatted IAS; and the remaining 84 acres had been used for harvesting, sowing, and mowing flower meadows.[49]

With the success of the pilot project, the project team is currently working to expand to other locations.[50]

## Performance and Evaluation

This technique creates wetland habitat, strengthens ecological connectivity, and boosts biodiversity. By focusing on corridors that connect core habitat zones, it facilitates species movement, feeding, and reproduction. The first species to colonize the wetlands tend to be aquatic insects, and dragonflies. Then, as vegetation continues to colonize the edge, amphibians arrive with birds following shortly behind.[51] Furthermore, if the ponds are laid out in a chain and are not too far from one another, species can use them as stepping stones or springboards when exploring new terrain.[52]

When compared to conventional landscape management techniques for reducing fuel under power lines, these techniques are estimated to be 1.4 to 3.9 times less costly over 30 years. This is despite a fairly significant initial investment.[53] The specific technique of creating ponds is a fairly quick and inexpensive process. A typical 1,000 square foot pond takes only three to five hours to dig and has an average cost of around $350. This cost can be further reduced if machinery is already onsite, if a number of ponds are being dug, if the terrain is flat and free of obstructions, and if the contractor is experienced.[54]

This technique also can transform overgrown parcels full of woody plants toward meadow and wetland-like landscapes with fewer IAS. This lower-profile vegetation is in an area that is highly vulnerable to ignition, helping reduce the risk of wildfire events. Furthermore, by maintaining large swaths of landscape below the power lines, linear fuel-breaks are created; these can help firefighters slow wildfires that are moving perpendicular to the corridor.

It also supports local shepherds, farmers, and breeders by providing new pasture space for their herbivores. It is also a great way to raise public awareness about habitat loss and wildfire risk. The technique is very visible to the public and is often transferable to other linear infrastructural networks such as railways, roads, and gas lines. The project shows how linear corridors like this can be multi-functional and can serve the needs of many stakeholders – from hunters to farmers and wildlife conservationists.[55]

## Challenges and Future Research

One challenge is that the safety corridor of the HV power lines may be owned by a number of people, resulting in a large number of project stakeholders. Thus, more communication and negotiation may be required to reach consensus.[56]

Another challenge is that this technique can pose a threat for TSOs to safely maintain the electrical equipment. With active grazing areas and designated habitat zones, the number of access points and service roads to this infrastructure may be limited. Additionally, while there are long-term cost savings associated with the project, there is often a significant initial investment that TSOs must overcome.[57]

The last challenge relates to long-term planning and how to adaptively manage these new landscapes within the safety corridors of the linear infrastructural networks. In the case of Drôme, funding was secured to ensure maintenance and monitoring efforts. But this is a relatively small site. With future projects, especially those with large sites, a long-term management plan for annual efforts will need to be prioritized.

1 mi

Adelaide

Cleland National Park

Cyborg Landscaping

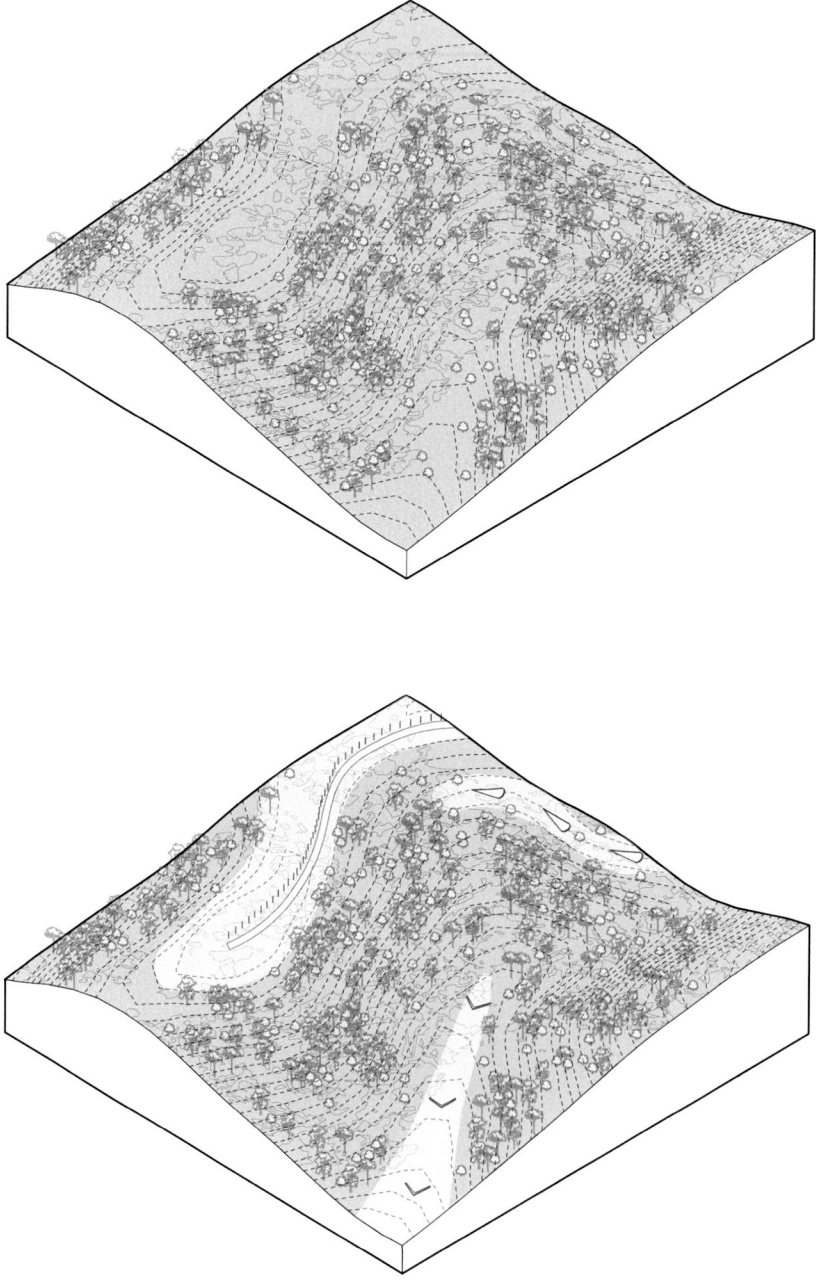

# 15

## *Cyborg Landscaping*

| | |
|---|---|
| Description | Embedding the landscape with sensors to jumpstart adaptation processes |
| Example Location | Adelaide Hills, Australia |
| Size | 2,500 acres |
| Implementer | Jordan Duke (conceptual proposal) |
| Team | Liat Margolis |

## Technique Overview

In January of 2015, the Sampson Flat bushfires ignited outside of Adelaide, Australia and went on to burn over 50,000 acres. While bushfires are a critical part of this landscape and positively bolster regional biodiversity, recent fires like this one have become more destructive compared to those in the past. Many attribute this shift to the effects of European colonization, climate change, invasive alien species (IAS) and growing development in the wildland-urban interface (WUI). In an effort to minimize damage to nearby structures and infrastructure, fire personnel in the region primarily rely on fire suppression techniques and traditional fuel reduction activities.[58]

In 2016, a Landscape Architecture masters student at the University of Toronto, Jordan Duke, sought to offer an alternative approach to managing wildfire for the region through her thesis work entitled "The Digital & The Wild: Mitigating Wildfire Risk Through Landscape Adaptations." With the project, she explored how real-time data monitoring could catalyze a range of ecological processes to increase landscape adaptation in the face of mounting wildfire threats.[59]

The project focuses on Cleland National Park located in the Mount Lofty Ranges just outside of Adelaide. The park is adjacent to a dense residential neighborhood, located on the edge of town. Prior to receiving its national park status, the landscape of Cleland had been cleared for timber production and farming. Currently, though, it is being regenerated through active restoration efforts. It features a range of ecological communities including riparian zones, bogs, heathlands, and forested areas.[60]

## Planning and Design Process

The proposed project has both short-term and long-term actions. Short-term, a fleet of environmental sensors embedded in the landscape provides real-time data on fire weather including slope conditions, soil moisture, atmospheric humidity, temperature, wind direction, and vegetative conditions. This information gives landscape managers and fire personnel a hyper-local and real-time perspective of the landscape, allowing them to better

Figure 5.14 (*Previous*) Aerial of Adelaide Hills. Source: Google Earth 2022.

Figure 5.15 (*Left*) A view of Cleland National Park (top) and the same view with weather modifiers, erosion accelerants, and artificial watering holes to catalyze change across the landscape (bottom).

understand the various conditions across the whole park. If any of the data being collected reaches a critical fire-danger level, they can automatically trigger devices to increase moisture in the soil and air. Longer-term, the same devices can catalyze larger wildfire adaptations for the landscape.

There are three primary devices driving the project.[61] The first is a weather modifier that is deployed along key ridgelines and valleys. When sensors detect potential fire weather, the modifiers immediately release moisture into the air to help slow or reduce the spread of an event. Longer-term, the modifiers can disperse seeds for fire-resistant plants like the Mediterranean cypress which is slow to ignite and can catch embers. The second device is an erosion accelerant strategically located in wet valleys. Over time, these capture water, which can be used for recreation, habitat, and firefighting purposes. When sensors detect fire conditions, the devices release water downslope, taking out fire-susceptible vegetation. Longer-term, weirs allow only clay particles to pass through to the valley, creating a wide firebreak for slowing or stopping fires and firefighting purposes. The third device is an artificial watering hole. These are deployed across the park, along key faunal routes and can be turned on and off based on fuel reduction needs. If parts of the park could benefit from foraging, managers can activate the watering hole and fill it with water, attracting animals who will munch on nearby vegetation. They also serve as key refuge areas during wildfire events.[62]

## Performance and Evaluation

This project capitalizes on inexpensive and easily deployable sensors that are already well used in the field of wildfire management to catalyze change across the landscape.

Also, instead of focusing on control, the project focuses instead on choreographing potential futures for the landscape by letting changes in landscape conditions dictate potential trajectories.

In addition, by having access to real-time data, it is possible to create effective feedback loops and to continuously adjust management based on current conditions. These data also help to shorten emergency response times to potential wildfire events.

Lastly, these devices not only monitor conditions and catalyze landscape changes, they also render these processes visible to the general public to raise awareness. This project, in turn, has the potential to shift attitudes and perspectives about the positive role of wildfire on fire-adapted landscapes.

## Challenges and Future Research

While there is much promise in the idea of cyborg landscaping, the technique faces a range of logistical hurdles.

First, while sensors themselves are inexpensive and fairly easy to deploy across the landscape, the three devices they are linked to may not be as straightforward or cost-effective to implement. Thus, they are in need of extensive testing before deployment can occur. For example, the erosion accelerants may not have much water in them during fire season, and if they do, the erosion process may actually initiate a weed infestation, which is a fire risk in and of itself.

In addition, over time, the sensors and their associated devices may burn or break or there may be extensive network interruptions during wildfire events.

1 mi

Rogue Siskiyou National Forest

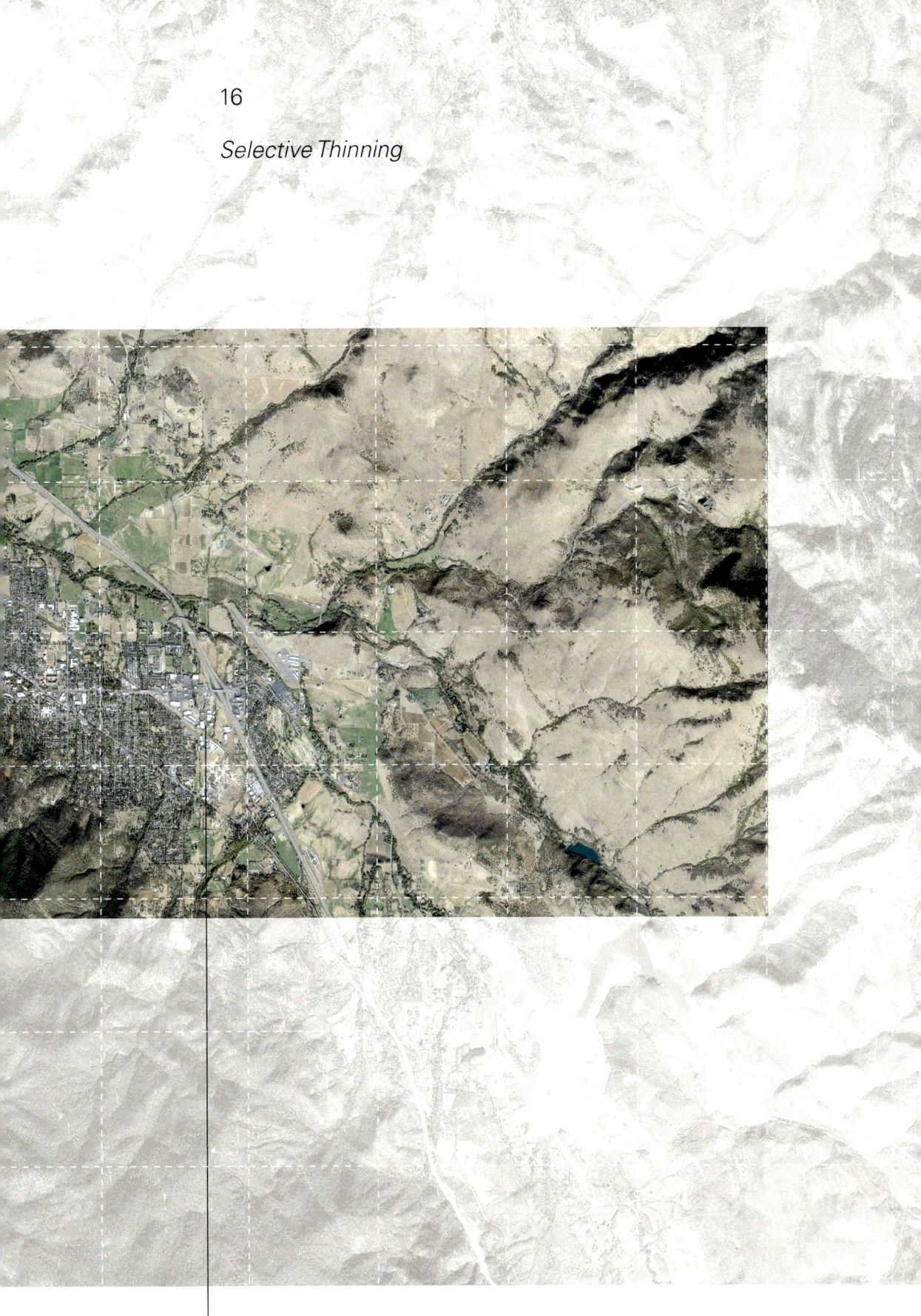

Ashland

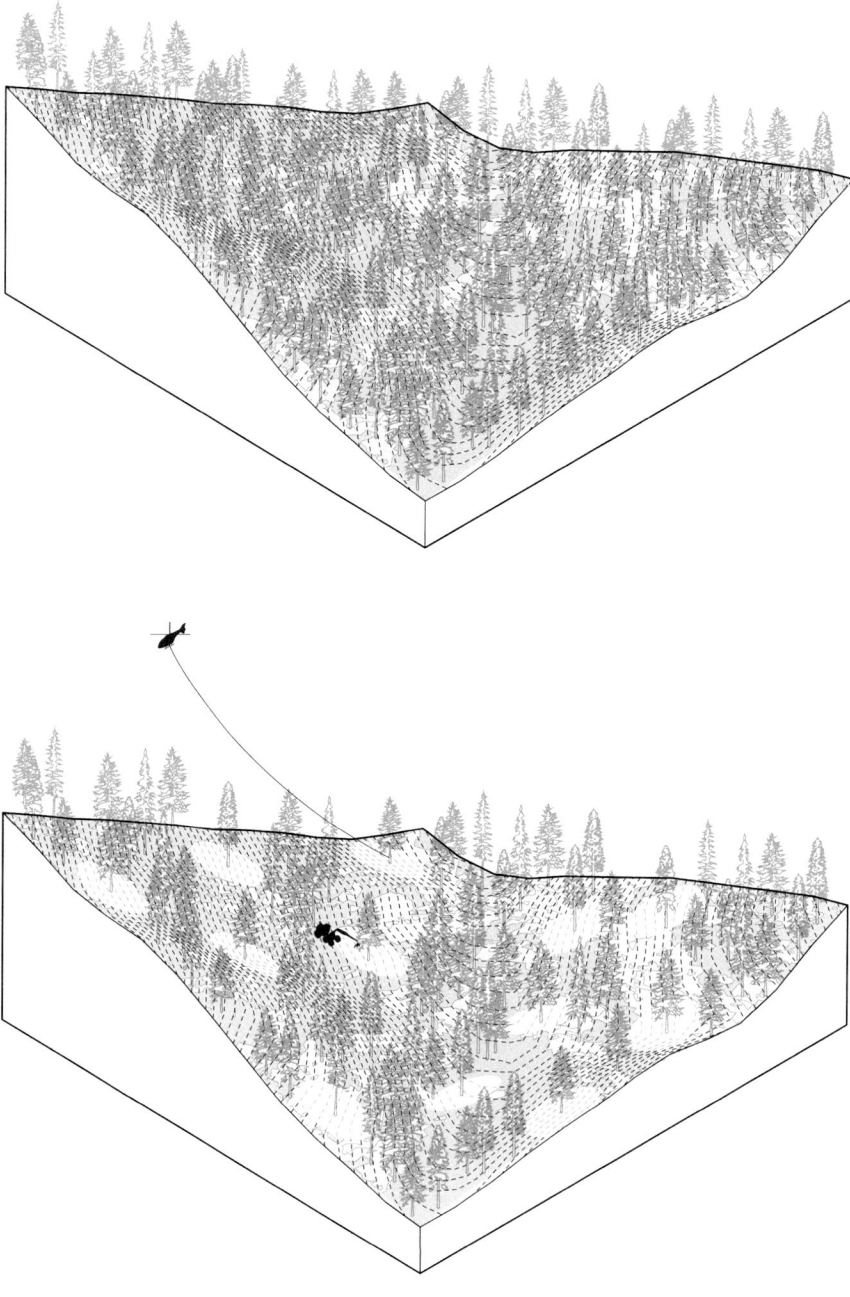

## 16

*Selective Thinning*

| | |
|---|---|
| Description | Reducing the number of trees per acre and economizing the by-products |
| Example Location | Ashland, Oregon |
| Size | 13,264 acres |
| Implementer | U.S. Forest Service |
| Team | The Nature Conservancy, Lomakatsi Restoration Project, City of Ashland, USDA Natural Resources Conservation Service, Jackson Soil and Water Conservation District, U.S. Fish and Wildlife Service, Oregon Department of Forestry, Oregon Watershed Enhancement Board, Bureau of Land Management, US National Park Service, Southern Oregon University, National Park Service, Klamath Bird Observatory, Southern Oregon Forest Restoration Collaborative, Oregon State University Forestry and Natural Resources Extension, Klamath-Siskiyou Wild, Private Forestland Owners. |

## Technique Overview

In 2010, the USFS entered into a ten-year agreement with The Nature Conservancy (TNC), the Lomakatsi Restoration Project (LRP), and the City of Ashland to implement the Ashland Forest Resiliency (AFR) stewardship project. With the project, TNC focused on ecological analysis and monitoring, LRP focused on workforce development and physical restoration work, and the City of Ashland focused on community engagement.[63] Over time, more stakeholders got involved to aid with the project.

In general, the project had four primary steps. After defining objectives for the project, the project team modeled a range of potential treatments for the forest and developed performance metrics. Then the team ran three scenarios to better understand which one would be most successful in reducing risk: a no-treatment scenario, a public land treatment scenario, and a public and private land treatment scenario. In the end, the last scenario was selected for the project and a checkerboard of federal and industrial logging land was used for treatment.[64]

With the project, the team focused their efforts on four goals: (1) protect water resources, threatened species, late successional habitat, human lives and property, legacy trees, and sensitive ecological systems, (2) <u>fuel reduction</u>, (3) reduce the potential for <u>crown fires</u>, and (4) bolster the health of forests in the face of wildfire. In general, the team aimed to mimic historical disturbances in the forest that were more frequent but were low-to-moderate severity.[65]

Figure 5.16 (*Previous*) Aerial of Ashland. Source: Google Earth 2022.

Figure 5.17 (*Left*) A view of the forest before treatment (top) and the same view with radial thinning, thinning-from-below, and non-commercial thinning to reduce the density of the canopy (bottom).

## Planning and Design Process

To determine detailed treatment prescriptions, the team first analyzed <u>management blocks</u> by looking at existing soil stability, soil resource protection, forest health, wildlife, snags and coarse woody material, hydrology, botanical resources, native grass seeding, noxious weeds, cultural resources, and logging systems. With this information, they identified treatments for each block.

One common treatment, called <u>radial thinning</u>, is the removal of all trees 15–50' around legacy trees to create vertical and horizontal <u>fuelbreaks</u> around the most highly prized trees in the forest. Another treatment, called <u>thinning-from-below</u>, is a stand-level removal of trees in the understory that would likely die due to the density of the forest. Wood from both of these first two treatments could be sold to mills. <u>Non-commercial thinning</u>, is a treatment entailing stand-level removal of smaller trees or vegetation unsuitable for the timber industry. Another treatment, called <u>piling and burning</u>, creates strategically located <u>fuel</u> piles for wildlife habitat.[66]

The <u>thinning</u> work was done aerially through the use of helicopters (at $876/acre) or on-the-ground through the use of low-impact tractors (at $274/acre). The <u>prescribed burning</u> work was done on-the-ground (at $650 to $750/acre).[67]

To effectively manage the project, the team set up a multi-party monitoring system focused on administrative monitoring for tracking money and time, implementation monitoring for tracking treatment procedures, and efficacy monitoring for tracking ecological and social outcomes.[68]

## Performance and Evaluation

Over the course of ten years, the team behind the AFR stewardship project treated 13,264 acres of public and private land in the Rogue Siskiyou National Forest, which is approximately 25% of the total forested acreage in the project area. Of this acreage, approximately 90% was treated via <u>non-commercial thinning</u>, 22% was treated via commercial <u>thinning</u>, and 11% was treated via <u>prescribed burns</u>.[69]

The project had many positive ecological outcomes. According to the project team, the thinning efforts did not negatively impact two species of concern – the northern spotted owl and the Pacific fisher – sensitive soils were protected, erosion was not accelerated, water delivery was increased in streams, and canopy fire potential was reduced as well as <u>wildfire suppression</u> difficulty. Furthermore, the thinning efforts created an open canopy, giving remaining trees more resources to thrive.[70]

To discover these findings, the team carefully tracked the progress of the project through a transparent and robust, data-driven monitoring platform. In doing this, they were able to learn from their own work and adapt their management protocols, create a baseline for tracking future shifts in the watershed, and build trust with stakeholders.[71]

Over the course of the project, public support for thinning and <u>prescribed burns</u> grew, partially due to the recent occurrence of wildfire events in the region, but also likely due to the outreach efforts associated with the AFR stewardship project.[72]

Economically, the team received approximately $6 million which was reinvested in the project. The sale of over 100,000 trees to local mills created 14 million board feet, which could build over one-thousand 2,000 square-foot houses.[73] Furthermore, the project supported a number of seasonal restoration-based jobs and supported sales at local businesses.[74]

Early on in the development of the project, it was decided that an "All Lands, All Hands" attitude toward managing the landscape outside of Ashland was necessary to bring about meaningful change.[75] In the end, this collaborative approach supported co-investment and buy-in from partners and paved the way for future stewardship actions.[76]

Based on the success of the AFR stewardship project, a regional forest restoration effort is being launched for the larger Rogue Basin. The project is considering landscape-scale goals focused on safety, habitat restoration, and landscape adaptation.[77]

## Challenges and Future Research

Given the scale and number of stakeholders involved in the AFR stewardship project, coordination and management was a significant challenge. The project had many goals that needed to be addressed including timber extraction, water quality, air quality, the protection of vulnerable neighborhoods, species conservation, climate adaptation, and the protection of indigenous resources.[78] Therefore, it was crucial that all stakeholders involved in the project were on the same page about the scope and procedures involved in achieving these goals to maintain a consistent vision.[79]

Another challenge of the project was land ownership and accessing property for management. The land outside of Ashland is a checkerboard of public and private ownership, requiring an extensive approvals process.[80]

With the project, it was also difficult to measure all of the co-benefits associated with treatment efforts. These benefits included: an increase in biodiversity, climate adaptation, fire resistance, carbon sequestration and a decrease in water stress.[81]

Building sufficient capacity to conduct the restoration work was another hurdle the team faced. This kind of project required a diverse workforce with skills to operate hand tools, machinery, analyze and map sites, conduct prescribed burns, suppress fires, engage the public, and monitor progress. Thus, identifying capable team members to carry through with the work and providing sufficient training became crucial elements of the project.[82]

Funding for the project came from a wide range of sources, including the partners of the project and a number of federal sources. To appropriately replicate and scale this project, long-term funding should be secured. One idea for doing this is to create an institution focused on building community adaptation in fire-prone areas.[83]

## Selective Thinning Interview

Figure 5.18 Kerry Metlen, Forest Ecologist at The Nature Conservancy.

DESIGN BY FIRE: To start, could you tell me a little bit about yourself, your role at The Nature Conservancy, and how you got involved with the Ashland Forest Resiliency (AFR) project.

KERRY METLEN: I was born and raised in Northeastern Oregon, and from an early age knew I wanted to be in science. I ended up working for the Forest Service Research Station studying forest and fire there in the late 90s and working on some fire and trail crews and getting to see some of this stuff firsthand. But what I really wanted was to be in research, so I went back to school and got my master's and my PhD at the University of Montana. I had the great opportunity to be part of a visionary fire and fire surrogate study, where we took an experimental approach to applying what we thought were appropriate ecological thinning and prescribed fire treatments and then tracking the understory over time and, that's really what kickstarted my academic career. When it was time to graduate and get a job, I'd always wanted to work either in research or for The Nature Conservancy, because I really value how the work we do gets applied to better stewardship. I was fortunate enough to get a job coordinating the multi-party monitoring for the Ashland Forest Resiliency project, which afforded a great opportunity to engage with the community. And through that we also had the opportunity to start bolstering our understanding about fire ecology in Southwestern Oregon. I was really stoked about the opportunity to catalog fire scars, and then publish a paper on fire histories of Southwestern Oregon showing the really strong connection to fire regimes of the Klamath mountains, and then deliver that to our collaborative groups to inform what kinds of prescription targets that we should be shooting for. For example, should restoration treatments focus on small units along roads or do these need to be more landscape scale? And then eventually, as AFR became a successful model of working on federal land with stewardship agreements and in partnerships, we were able to expand to management toward shared objectives on city land and private land. We wanted to grow that model so, working with the Southern Oregon Forest Restoration Collaborative, we ended up writing a cohesive forest restoration strategy for southwestern Oregon where we identified some possible objectives at a landscape scale. It includes fire safety for communities, but also restoring the habitat of the Northern Spotted Owl and promoting landscape adaptation.

Now what's coming to fruition is that we're expanding from Ashland to funded work from the Oregon Watershed Enhancement Board to start seeding these ideas across the Rogue Basin. There's a strong interest from NRCS in growing that effort to private land, and even philanthropic funding that's also coming together to help basically do a lot of work really emphasizing on communities, but also because we have this landscape perspective I think we're going to be successful in transforming landscapes in a way that has to happen. My sense is also that society's perception is shifting along with us. People are beginning to get on board.

DF: In terms of community perception, do you feel like there has been a shift over time where people are more accepting of this kind of work?

KM: Yes. And I know that because of the conversations I have, but also because we have a local study that has shown this significant shift in understanding of the need for landscape scale forest resiliency treatments. Over the course of the AFR project there's been an increase in support for implementing ecological restoration treatments at a landscape scale and for cutting trees of sizes supported by science. In general, I would say we were successful in our outreach and engagement efforts. Now, this increase in support correlates with multiple large fire and smoke events, so the shift in public perception comes largely because of contemporary events, not entirely because we did the work. While many people may not have heard of our project specifically, they usually support the goals of a project like AFR, which is still a great success.

DF: Is there interest coming from other communities, either in Oregon or California or elsewhere, in trying to use this framework and expand it or replicate it elsewhere?

KM: Definitely! AFR didn't happen in a bubble. There are other collaborative groups like in Bend, with the Deschutes Forest Project, and then the Blue Mountain Forest Partners. There are groups doing very similar work and we come together, and we share ideas from time to time. I'd say there is a lot of interest. As far as what some pointers are for these groups, I think it's really important to describe a shared vision. We need to come together as a community to discuss what are the things we value, what are the threats to those things, and how do we address those threats, and then we need to be very coherent about that vision. That can successfully lead to funding opportunities, and with funding comes the development of an organization with no rules about how you're going to play together and that provides the framework for success.

1 mi          Table Mountain National Park          Tokai Park

*Invasive Hacking*

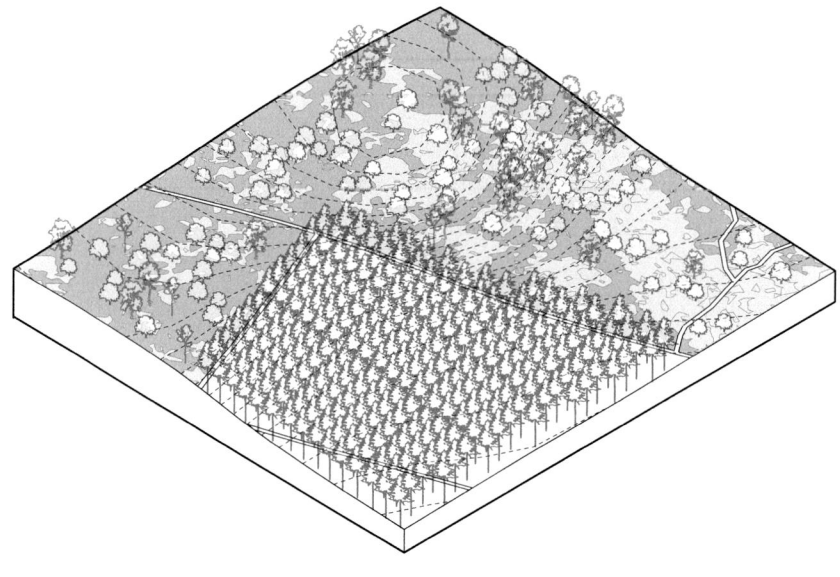

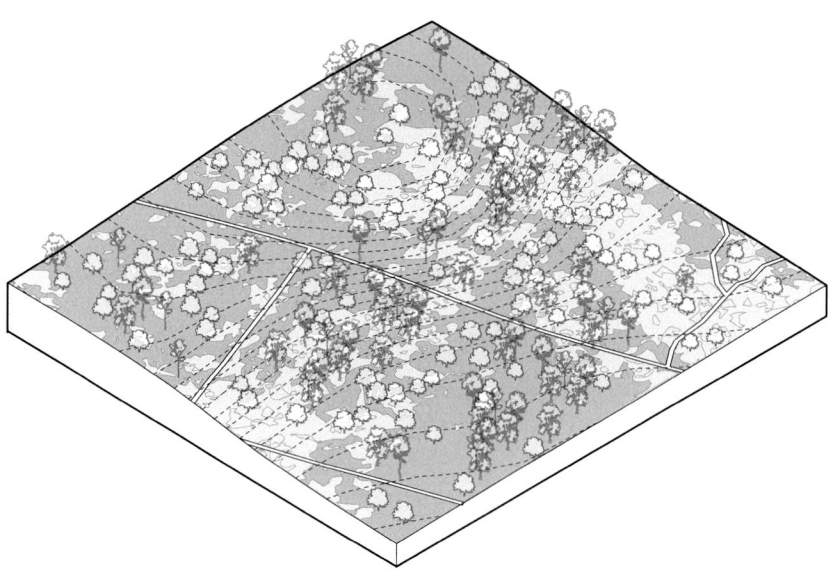

17

*Invasive Hacking*

| | |
|---|---|
| Description | Removing <u>invasive alien species</u> to promote native, fire-adapted fynbos habitat |
| Example Location | Cape Town, South Africa |
| Size | 1,400 acres |
| Implementer | Friends of Tokai Park (FoTP) |
| Team | South African National Biodiversity Institute (SANBI), South African National Parks (SANParks), Table Mountain National Park (TMNP), Working for Water (WfW), The Wildlife and Environment Society of South Africa (WESSA), City of Cape Town, local neighborhood watches |

## Technique Overview

Tokai Park is located on the eastern side of Table Mountain in the Cape Floristic Region (CFR). The western side of the park connects up with the TMNP and the eastern side is flanked by suburban development. It is one of the last areas in the region that connects lowland fynbos to mountain fynbos, linking critical ecological processes.

Since the early 1900s, the property at Tokai has been used for commercial forestry, with tree plantations covering much of the landscape. It has undergone three cycles of planting and harvesting, with some timber supporting WWI and WWII efforts, and has also been the site of many experiments.[84] Throughout this time, residents of Cape Town have used the plantations for a range of recreational activities including walking, hiking, horseback riding, mountain biking, and picnicking.[85]

In 2005, the management of Tokai Park was handed over to South African National Parks (SANParks) and the commercial forestry company was given 20 years to harvest the last 1,500 acres of tree plantations. To aid in the rehabilitation of this post-plantation landscape, SANParks developed a management framework that established a series of biodiversity goals including: restoring threatened vegetation types, protecting sensitive zones, linking ecological corridors, maintaining habitat for indigenous animals, managing IAS, and handling fire.[86] To aid with these restoration and conservation efforts, an environmental conservation organization called Friends of Tokai Park (FoTP) formed and has been working alongside SANParks.

Figure 5.19 (*Previous*) Aerial of Cape Town. Source: Google Earth 2022.

Figure 5.20 (*Left*) A view of a commercial forestry plantation (top) and the same view after removing the invasive alien species to promote native, fire-adapted fynbos habitat (bottom).

## Planning and Design Process

A number of roads cut across Tokai Park, breaking the space up into a series of <u>management blocks</u> that have been used to harvest timber. Following the final harvest, a series of restoration efforts tend to follow.[87]

First, SANParks conducts prescribed burns across the blocks as part of their continuous fire management of TMNP. To conduct these burns, SANParks fire managers work closely with FoTP to develop burn plans. And while fynbos historically burn during the hottest months of the year, these burns tend to happen during the cooler months to help with containment.

Once the blocks have been burned, passive or active restoration techniques are used to recover indigenous flora and fauna.[88] If the fynbos seed bank in the soil is intact but dormant, there is potential for passive restoration and spontaneous plant succession, but if it is gone, there is a need for active restoration. Active restoration involves sowing seed and planting starts in the winter months.[89] Recent studies on this topic point to a concept called the restoration threshold – a point in which the number and density of recovered species from a seed bank significantly decreases due to the age of the plantation.[90]

During this process of passive and active restoration, FoTP surveys each block for IAS, especially wattle, acacias, and gums which can quickly colonize the landscape. To deal with these species, FoTP has organized year-round, monthly hack events as an IAS management strategy.[91] Once identified, IAS are either removed with hand saws, loppers, or chain saws. Herbicide is also typically applied to stumps to prevent regrowth.[92]

In Tokai Park, hack events are organized differently for the lower part of the park compared to the upper part. The terrain of the lower part is generally flat and comprises Cape Flats Sand Fynbos, the most threatened type of vegetation in the park.[93] It is also an area that has been heavily impacted and fragmented. Typically, hack events in this part of the park last 2–3 hours with a small group of 3–9 volunteers.[94] The upper part of the park has steeper, less accessible slopes, and comprises Peninsula Granite Fynbos, Peninsula Sandstone Fynbos, and small areas of Afromontane forest.[95] Due to the terrain and the high density of IAS, this part of the park requires more planning and experienced volunteers. Hack events here also last 2–3 hours at a time, with 6–19 volunteers working together at once.[96]

Throughout these efforts, volunteers map and track the vegetation in each block by posting observations on iNaturalist.[97] FoTP also uses machine learning algorithms to detect park land cover from above.[98]

In the future, FoTP hopes to officially launch an adopt-a-plot program to expand their volunteer base even more. With this program, they hope to engage schools, companies, and other organizations to get trained in hacking, help remove IAS in one area of the reserve, and monitor their progress.

## Performance and Evaluation

The restoration efforts in the lower part of Tokai Park are succeeding – hundreds of indigenous plant species have returned along with associated animal species. The project has also strengthened the larger ecological network of the region by making a key connection between the lowland and upland fynbos. This success bodes well for other patches of highly threatened Cape Flats Sand Fynbos in the region.[99]

The data-driven approach behind FoTP's work allows the organization to properly prioritize and plan restoration efforts across the treatment blocks. Since launching their monitoring program, volunteers have made over 12,000 observations of over 1,000 species in the park.[100]

FoTP's focus on citizen science and the active recruitment of community members has increased awareness about the benefits of conserving and restoring fynbos, the threats of IAS, identification procedures, removal techniques, and ways to report observations and progress. In participating, they can also build social capital by meeting other volunteers and can engage in challenging physical activity.[101]

## Challenges and Future Research

While the efforts at Tokai Park have succeeded on many fronts, the initiative faces a range of challenges, including access to long-term funding. Currently FoTP is a volunteer-based organization and it is often difficult to maintain and attract new members. Furthermore, it is expensive and time-consuming to remove IAS since they establish quickly and it is necessary to closely monitor the landscape to coordinate removal efforts.[102] If left for too long, many species can no longer be removed by hand and must be removed mechanically. Furthermore, it takes time to train volunteers to properly identify species and to safely access steep slopes.[103]

Another challenge concerns the coordination between organizations involved in the management of Tokai Park's landscape. In order to efficiently and appropriately restore and manage the fynbos, organizations should be in close communication regarding the timing and extent of prescribed burns and clearing events so as to not be duplicative or counter-productive to past or future efforts.

Public perceptions about Tokai Park is another hurdle for the project. Some community members feel as though current restoration efforts do not adequately address the needs of those who live nearby. A primary concern for this group is the loss of shaded recreational space if the plantation is removed.[104] There are also concerns about crime, as some believe the vegetative structure of fynbos can make the park more dangerous. Lastly, there are fears about the relationship between fynbos and fire at the urban edge. Some residents push for frequent fuel reduction burns to reduce risk (at intervals not conducive for fynbos growth) while others push solely for fire suppression efforts.

There are a range of logistical issues related to current restoration efforts. First, there is the potential for cross contamination. For example, transplanted seedlings may have unknown genetic material in them and garden tools used for hacking or planting may have unknown soil microbes on them. Furthermore, with active restoration, the sourcing of local plants and seeds may be difficult. Lastly, given that the lower part of Tokai Park is one of the last remnants of Cape Flats Sand Fynbos in the region, the work being done by FoTP cannot easily be replicated. The team has identified trends, but all of their findings are anecdotal.[105]

*Invasive Hacking Interview*

Figure 5.21 Alanna Rebelo, Postdoctoral Researcher in Conservation Ecology and secretary of the Friends of Tokai Park committee.

DESIGN BY FIRE: Why does Tokai Park matter globally?

ALANNA REBELO: Tokai Park is a 1,500 acre wing of Table Mountain National Park, one of South Africa's UNESCO World Heritage Sites. This small peninsula of Fynbos extends into the urban matrix and is one of the only corridors connecting the mountains to the lowlands. The park contains four vegetation types: Cape Flats Sand Fynbos, Peninsula Granite Fynbos (both critically endangered), Sandstone Fynbos and some small patches of Afromontane Forest. Two of these vegetation types are critically endangered, with Cape Flats Sand Fynbos being less than 1% conserved. Plantation forestry is the reason that this small parcel of land was not developed, leasing land for forestry operations for over a century. Surprisingly, the fynbos seedbanks under these pine plantations have persisted and some very rare endemics have re-emerged. Today there are more than 550 native plant species in the Cape Flats Sand Fynbos, many of these rare endemics. There are 42 IUCN threatened plant species at Tokai Park alone.

DF: Can you talk about the current forestry operation at Tokai Park?

AR: There are just a few blocks of pine plantation left, on the Cape Flats Sand Fynbos, all due to be harvested by 2024, and the park authorities are supposed to be returning this land to conservation thereafter (fynbos shrubland). The forestry company that used to be in the area left behind a massive legacy of alien tree infestations in the park, particularly on the mountains in the Peninsula Granite Fynbos. These alien trees are a major threat to biodiversity, have negative impacts on water availability as well as increasing the fire danger of the area. Following fires in 2016, the forestry company tried to harvest all their pines since they had to bring in machinery anyway to harvest the burnt pines. It was most economical to do everything at once. But some community members fought against this, getting a court interdict to stop this company harvesting its own trees, because they wanted to retain the pines.

DF: Can you tell me more about the events that your group hosts?

AR: Our main activity is alien hacks. Usually a small group of people meet every other week or so, and tackle one site at a time. In 2020, we took many school groups to a site at Tokai Park which had been clear cut and burned, and there were all these wattle tree seedlings coming up. If we had left the seedlings another year, that whole area would have been totally infested with trees, but we got that whole block cleared over the course of the year. What's been much more difficult is clearing up on the slopes, because you've got to hike to reach the areas and it's very intense. Instead we've identified a few priority areas based on threatened species that we do our best to keep cleared. The hacking process is still running but we're hoping to get an adopt-a-plot program running and get more school groups involved as soon as the lockdown in South Africa eases up. The important thing about all this is that we give proper training, because we don't want big groups just running wild and pulling the wrong plants out.

Another dynamic that we have is with WfW which is a part of the South African Expanded Public Works Program dedicated to the issue of unemployment. WfW brings in a contractor who manages a group of 10–15 people to train and go out and do this invasive removal work. It's an important program, but we'd like to see their training brought up to a higher standard to avoid issues like pulling out native species, applying poison too liberally, and things like that.

DF: What do you see as the future vision and next steps for Friends of Tokai Park?

AR: Our challenge right now is building up our organization and bringing in more diversity, because a lot of young people aren't interested in being involved in committees like this. I am grateful for a lot of our members who are very active, some of the great events we've done, and the fact that we've been able to engage with the public more and more.

I also think our organization plays an important role as a watchdog, because we've seen a huge rise in land grabbing here in the Cape, where people will just claim land by squatting in nature reserves and other protected areas. So, we really serve as eyes on the ground for Tokai Park in that capacity, and we can also advocate against groups who might prefer to see the area developed for housing or even a theme park or garden rather than remaining as a wildland. We know that this area is critically important for biodiversity, and if we could secure some grants to create employed positions then we could really keep the momentum going in terms of preservation and management. Our ultimate vision is to see this area completely restored, and we know that people's perceptions and value systems can change so perhaps seeing this area restored would allow people to fall in love with the area. It has happened before in other parts of the Cape, like Silvermine Nature Reserve. My wish is that people would learn to appreciate their natural heritage more.

1 mi

Biamanga National Park

*Refugia Expanding*

The Crossing

Bermagui

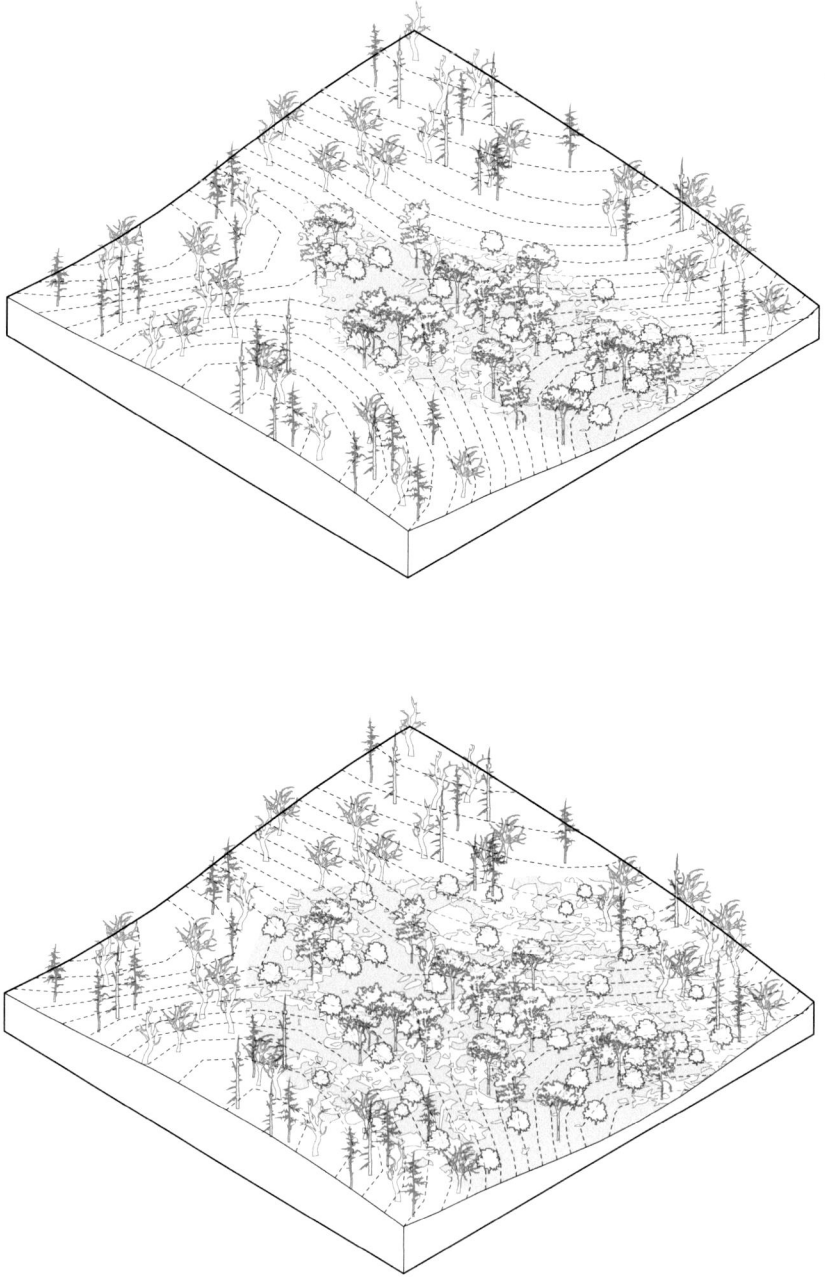

# 18

## *Refugia Expanding*

| | |
|---|---|
| Description | Enhancing <u>fire-resistant</u> native habitat remnants through weed control |
| Example Location | Bermagui, Australia |
| Size | 58 acres |
| Implementer | The Crossing |
| Team | Landcare Australia |

## Technique Overview

The 2019–2020 Australian <u>bushfire</u> season was deemed the "Black Summer" due to the number and intensity of wildfires that erupted over the course of the year. One of these fires, the Badja Forest Fire ignited in late December of 2019 and ultimately burned over 1,000,000 acres and destroyed over 1,400 structures.

Following the fire, some patches of the landscape were minimally burned or left untouched. These patches are often called <u>refugia</u>, residual islands, biological legacies, or fire shadows and are defined by a range of attributes. Some refugia can appear after one fire, then disappear during the next – these are called <u>ephemeral refugia</u>. While some can withstand multiple fires over a prolonged time – these are called <u>persistent refugia</u>. Also, they can range in size from .025 acres to 250 acres. Typically, cooler and more moist areas of the landscape that support late-successional species with fire-resistant qualities can become refugia.[106]

In the peri-urban landscapes of Australia, citizens and local governments have increasingly focused on protecting and expanding fire <u>refugia</u> in an effort to regenerate fire-impacted bush habitat. One place where this is happening is at The Crossing, an environmental education camp that was severely impacted by the Badja Forest Fire.

## Planning and Design Process

At The Crossing, land managers are actively managing and enhancing fire <u>refugia</u> that are tucked into onsite gullies. Here, the landscape has escaped the effects of wildfire for 60–70 years and is significantly wetter and cooler than adjacent areas. For the last few years, employees of The Crossing have worked to expand the condition in these areas using the Bradley method.

Developed in the 1960s, the Bradley method was developed to ensure that refugia do not become overgrown with weeds following a major disturbance, like a wildfire. If weeds overtake a site, they can displace native plants by competing for light and altering soil nutrients, ultimately leading to habitat loss and an increased <u>fuel load</u>.[107]

There are two main principles driving the Bradley method. The first is that regeneration should happen with minimal disturbance. Instead of using heavy machinery which can exacerbate onsite disturbance, tending should happen by hand. Weeds should be carefully pulled, and soil should be immediately covered with mulch or leaf litter to deter future growth.[108] The second principle is that regeneration should happen slowly over time. It is advised to focus on small areas and to slowly move to new areas once there are signs of regeneration. Oftentimes, one or two people can effectively regenerate a site this way, instead of involving a large group.[109]

There are a number of steps involved in the Bradley method that can be replicated in different places by different groups. The first step is to do a plant survey of the refugia to understand what species currently exist on the site. This is important for identifying native plants that should be preserved, as well as invasive weeds that should be pulled. The process of regeneration should then start in the area of the refugia that has the least number of weeds. Here, weeds should be removed once or twice a year. After this area begins to regenerate, the team can move to a new zone with a larger percentage of weeds. Here, it is suggested that people focus on 12-foot-wide swatches of the site at a time, slowly removing all invasive plants. Throughout the course of the year, these swatches can extend in length if the regeneration process is successful and the native plants begin to form a dense groundcover. Over time, the team can begin to work in areas that are predominantly weeds. It is important to regenerate the edges of these zones first so that native vegetation can have a chance to take over. For these areas, it is suggested to work in six-foot-wide patches and to focus first on spot weeding – removing large weeds growing adjacent to native plants. The last step of the Bradley method is to record the regeneration process through written records and maps. This allows teams to better track their progress over time, so that they can adjust their management if necessary.[110]

## Performance and Evaluation

Refugia play an important ecological role during and after wildfire events. First, they serve as retreat zones for animals looking to shelter during active fires. This is especially important for those with limited mobility options. Shortly following a fire, these areas can become critical remnant habitat for individuals who lost their prime habitat due to wildfire. In this role, refugia can be a food source, a shelter from predators, and can serve as a stable location less vulnerable to post-fire issues like erosion and slides. Also, if areas around a refugia are severely burned in a fire, the refugia may help to disperse seeds and expand habitat into burn scars for decades following a fire.[111] This helps to protect the genetic diversity of these plant communities. Furthermore, the expansion of refugia can help make a landscape more resilient in the face of future wildfires, as refugia can naturally break up the activity of fire fronts.

Beyond ecological benefits, refugia expansion through the Bradley method also has potential for positive social outcomes. For instance, this process can improve the value of a landscape by improving the aesthetics and increasing amenity opportunities for the community. It can also help to protect historical and cultural sites of significance that might be vulnerable to bushfires. Furthermore, the act of carefully tending the landscape can build morale within a community following a bushfire. This process can also help to ultimately reduce long-term maintenance efforts.[112]

## Challenges and Future Research

While the process of regenerating refugia following a wildfire has many benefits, there are some logistical challenges. To begin, the Bradley method is intended to be slow and methodical – thus, regeneration may take a long time and require a lot of physical effort. Furthermore, extensive coordination might be needed at the local and regional scale to properly train and incentivize landowners and land managers. This will be necessary to catalyze change at a large scale.

Additionally, the Bradley method as initially developed back in the 1960s, may need to be modified based on new ecological knowledge. For instance, rather than promoting a "no disturbance" approach to these areas, it may be necessary to introduce some level of disturbance. This is especially important to jumpstart seed banks in the soil. The method may also need to be augmented with other techniques like the targeted application of herbicide, the construction of fences, and the planting of native plants and trees.[113] Lastly, it should be noted that some weedy species may play important ecological functions, and these should be considered before their removal. For instance, in the absence of native species, introduced blackberry plants can serve as critical habitat for native birds.

1 mi

Bloubergstrand

Blaauwberg Nature Preserve

19

*Block Burning*

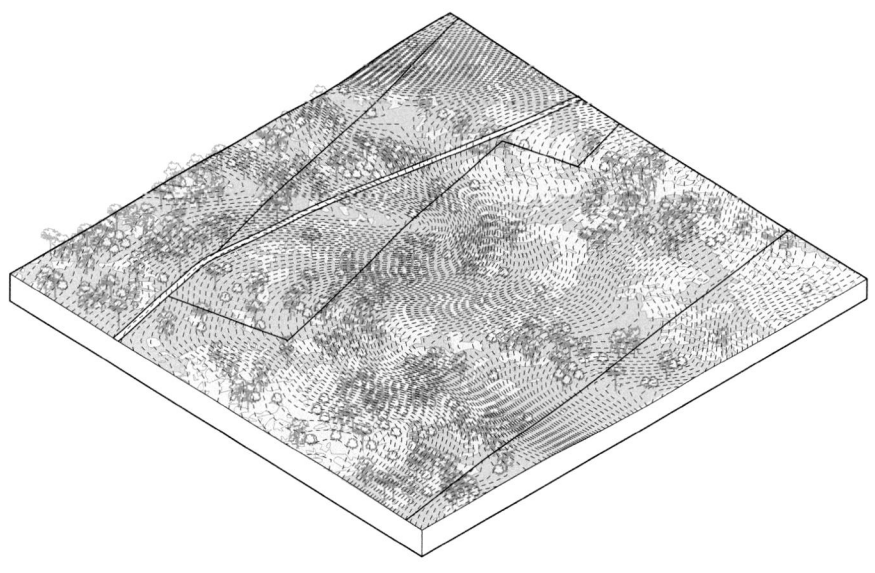

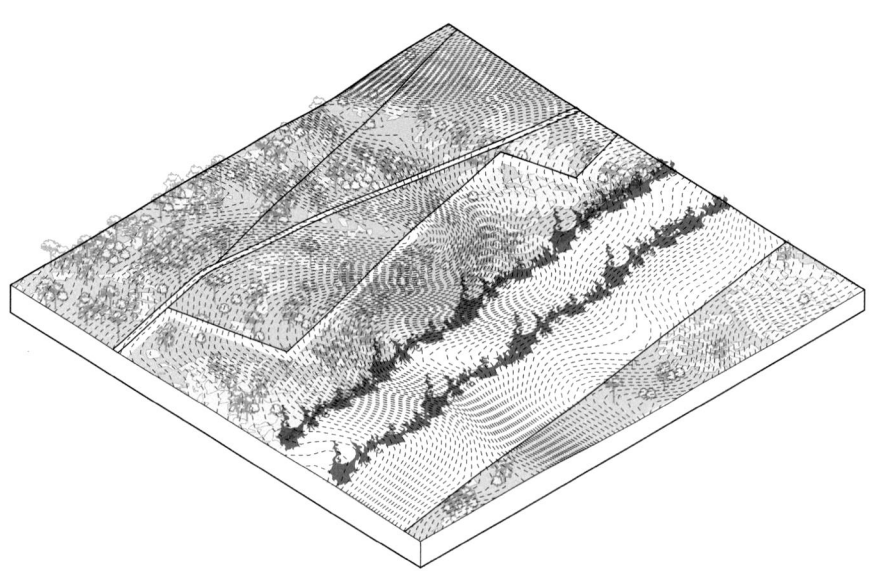

19

*Block Burning*

| | |
|---|---|
| Description | Prescribed burning done in management blocks on a rotational basis |
| Example Location | Cape Town, South Africa |
| Size | 3,700 acres |
| Implementer | Blaauwberg Nature Reserve |
| Team | Friends of the Blaauwberg Conservation Area, Millennium Seed Bank Project (MSB) of the Royal Botanic Gardens, Stellenbosch University and City of Cape Town |

## Technique Overview

In the late 1970s, there was a significant shift in how land managers in the CFR perceived fire. Due to an increased interest in fynbos ecology, land management practices pivoted from fire suppression to the reintroduction of fire through prescribed burning practices. At this time, ecologists and land managers began noticing a decline of charismatic fynbos plant species, they realized that periodic fires could actually help reduce water loss, and they understood the impossibility of total fire exclusion on the land.[114]

One of the primary prescribed burning techniques that emerged out of this shift was block burning. Like many other controlled burning techniques, block burns are set intentionally and are carefully monitored under controlled conditions to reduce fuel on a site. The technique typically involves a site that is broken down into management blocks of a similar size, shape, and age since the last burn. These blocks are then burned on a rotational basis in the late winter or early spring, when weather conditions are favorable.[115] The goal is to get a checkerboard-like effect with the blocks to slow down future wildfire events and prevent severe wildfires from negatively impacting the whole site.[116]

## Planning and Design Process

Block burning has been implemented in Blaauwberg Nature Preserve since 2013. The preserve, which was established in 2007, sits just north of Cape Town directly east of Big Bay Beach. It is about 3,700 acres in size and contains critically endangered vegetation. This vegetation, while degraded and largely taken over by introduced species such as acacia, is one of the last remaining patches of Cape Flats Sand Fynbos in the region; thus, the site has been identified as having significant conservation potential.[117]

To prep the site for block burning, land managers first divided the preserve into large management areas called fire management blocks (FMB). The shape and size of the FMBs were determined by the age of the vegetation as well as existing physical borders, including natural borders such as rocks or cliffs and built borders such as roads. These physical borders served as firebreaks, preventing the controlled burn from impacting other areas.[118] The blocks were then cleared of <u>invasive alien species</u> (IAS). Following this removal, the block burning began in early 2013, when weather conditions were favorable.

To begin the block burning, fire managers first lit a back fire on the far edge of the block perpendicular to the dominant wind direction. With this technique, the fire moved into the wind so the flames moved slowly and had short flame lengths; thus, it was easy to control. This backfire created extra protection for the existing fire break. Following the back fire, fire managers lit flank fires which ran parallel to the dominant wind direction. Once the back and flank fires burned the edges of the block, the fire managers lit a <u>head fire</u>. This fire burns with the dominant wind direction, spreading quickly with long flame lengths. Eventually, these three types of fires merged in the middle of the block and burned out. If the fire self-extinguished before burning the whole block, fire managers reignited to ensure a complete burn.[119]

## Performance and Evaluation

<u>Block burning</u> is especially effective for wildlands that are adjacent to or nearby highly populated urban areas. In these places, wildfires or less regulated intentional burns cannot happen for safety reasons.[120]

Furthermore, block burning can be a practical and efficient way to manage fire-prone and fire-adapted landscapes. Due to the simple geometry of the FMB system and its regular and predictable burn schedule, it is easy to keep track of what has been burned and needs to be burned every year. This systematic approach can also reduce management costs. Furthermore, some land managers have created priority ranking systems to decide when each block should burn. These systems balance factors such as site ecology, tourism, infrastructure, IAS, heritage issues, and nearby structures, creating a more balanced and objective decision-making process.[121]

Lastly, by using existing fire breaks (natural or constructed) as FMB boundaries, land managers can reduce the ecological impact on the site as well as the cost of constructing new lines.[122]

## Challenges and Future Research

One challenge with <u>block burning</u> is that the technique has a fairly homogenous effect on each block; there is just not enough spatial or temporal variability embedded in the process. Thus, it is less beneficial for creating a biodiverse habitat than other burning techniques like <u>patch-mosaic burning</u> (PMB).[123]

Another issue with the technique is the narrow window for when burns are permitted to happen. Ecologically, late summer-early autumn is the best time to burn fynbos vegetation, but due to the higher risk of <u>hostile fires</u> during this time, block burning is not allowed to happen during these months.[124] Instead, block burning can only happen during the winter and spring every year, and this window can be narrow if there is inclement weather.[125]

Furthermore, every block burn requires an extensive <u>burn plan</u> which takes time to develop and get approved. This administrative hurdle can lead to a backlog of burns from one year to the next and can ultimately result in a failure to create a checkerboard effect to adequately buffer the landscape from future fires.[126]

Lastly, many communities adjacent to areas treated with block burning are hesitant about the practice. People are concerned about fires burning outside of designated blocks, creating unnecessary air pollution, and changing the appearance of the landscape.[127] If there is enough opposition, management plans can fold.

1 mi

Bundanon

*Firestick Farming*

Shoalhaven River

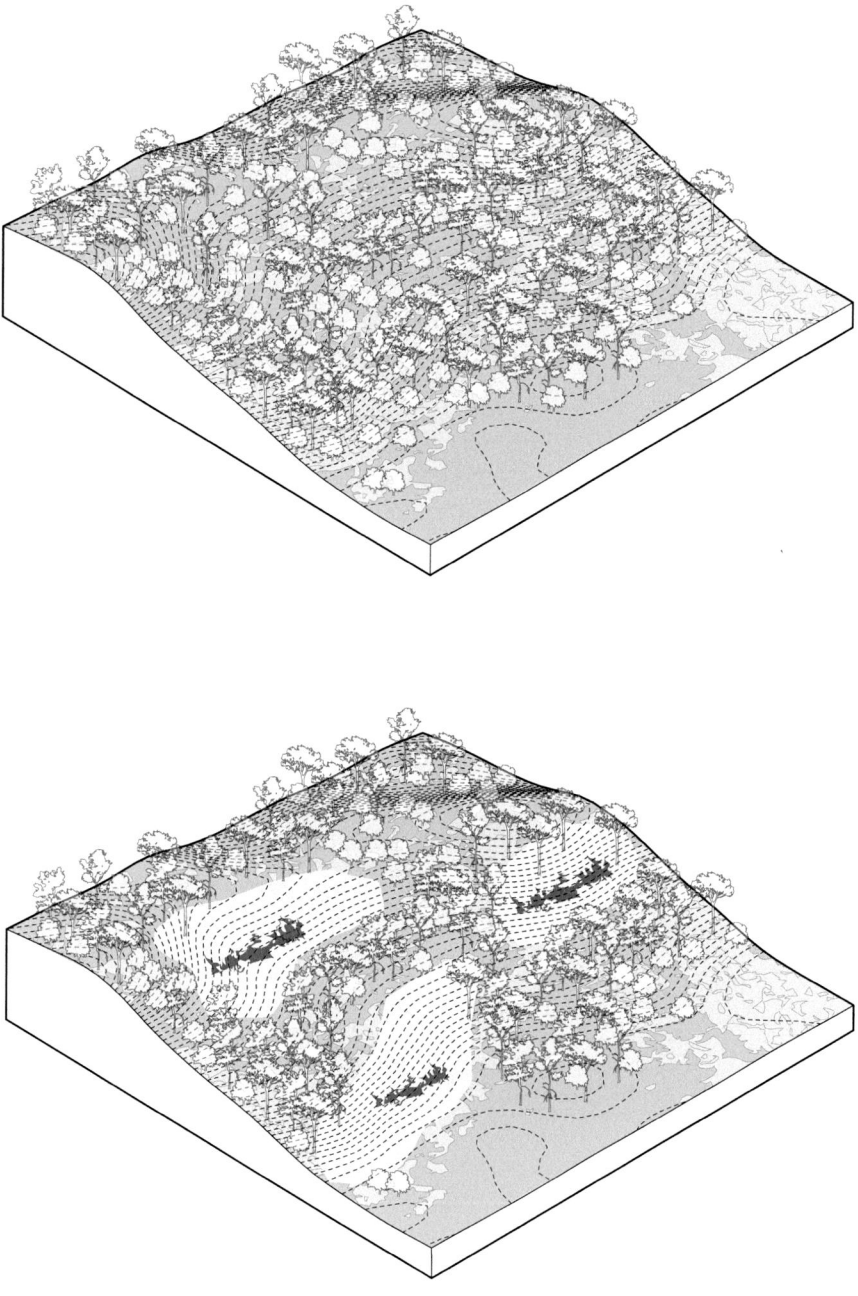

## Firestick Farming

| | |
|---|---|
| Description | Cultural burning to create mosaicked landscapes for harvesting and hunting |
| Example Location | Nowra, NSW, Australia |
| Size | 370 acres |
| Implementer | Firesticks Alliance |
| Team | Mudjingaalbaraga Firesticks Cape York Natural Resource Management, Mulong Productions, Bundanon Trust |

## Technique Overview

In 1969, Rhys Jones, an Australian archeologist, coined the term firestick farming to describe an indigenous practice of burning the landscape to manage resources. He claimed that when colonists first arrived in Australia, the landscape they saw was not "natural" (as they assumed) but, rather, had been managed with fire for thousands of years. He argued that this was an intentional and active practice, akin to farming, to clear ground for hunting and to regenerate the growth of edible plants.[128]

Today, firestick farming is also referred to as cultural burning, traditional burning, or aboriginal burning. This practice is often implemented on a small scale using single ignition point burns during the cooler months when the landscape has moisture to moderate burn intensity.[129] The practice is also very site-specific – indigenous knowledge of each landscape guides the burns. Once lit, these low-intensity fires move slowly across the ground, allowing practitioners to follow and guide them. David Bowman, Professor of Pyrogeography and Fire Science, describes them as "little fires tending the earth affectionately."[130] Over time, this practice results in finely mosaicked landscapes, a patchwork of habitats created by the fires' paths.

In recent years, there has been a growing interest in supporting cultural burning across Australia.[131] One organization that has been particularly successful in disseminating knowledge about the technique is the Firesticks Alliance Indigenous Corporation, a network of individuals focused on building community capacity for bringing fire back to the land. The alliance emerged in 2004 and has been hosting fire management workshops since 2008. Every year, the workshop travels to a new community and a new landscape.

Figure 5.26 (*Previous*) Aerial of Nowra. Source: Google Earth 2022.

Figure 5.27 (*Left*) A view of a heavily vegetation hillside (top) and the same view when using cultural burning practices to create mosaicked landscapes for indigenous harvesting and hunting (bottom).

## Planning and Design Process

In July 2018, the alliance gathered outside of Nowra, New South Wales, to hold their tenth annual workshop. Nearly 400 participants attended the four-day workshop. The first three days consisted of activities and lectures related to topics like ethnobotany, native plants, and traditional weaving techniques. Then, on the fourth day, participants conducted a cultural burn on the property. Unlike prescribed burns, these burns were done only with community members walking alongside the fire – without fire trucks, suppression tools, or "trained" firefighting experts.[132]

The cultural burn targeted two different habitats that had not experienced fire for over 35 years and had a significant amount of leaf debris covering the ground: the gum tree landscape, and the sand-ridge landscape. Without regular fire, the gum tree landscape cannot grow native grass in the understory and the sand-ridge landscape cannot build a robust seedbank in the soil and provide indigenous people with medicinal plants.

After carefully surveying both landscapes to ensure minimal faunal disturbance, the indigenous leaders lit a smoldering stick and ignited the ground with one, small spot fire. After 10 to 15 minutes, the fire began to creep across the ground with the participants following close behind. Over time, the participants ignited more spot fires, burning the landscape in a patch-like way. All of these fires burned slowly and low to the ground until they were extinguished 14 days later.[133]

## Performance and Evaluation

The act of firestick farming reduces fuel across the landscape in a way that is more ecologically and culturally sensitive than traditional large-scale prescribed burning techniques, while still providing protection from moderate bushfires.[134] Furthermore, by burning on a regular basis, these intentional burn sites are more likely to be eliminated from future wildfire events.[135]

This practice also promotes landscape regeneration by recycling older vegetation and increasing nutrient availability. The fine-grain landscape mosaic that results from firestick farming increases biodiversity and reduces habitat loss. This, in turn, has the potential to increase small-animal hunting productivity – facilitating a way to "farm" game.[136]

Lastly, firestick farming is an inclusive practice that foregrounds the knowledge of individuals who have historically been excluded from traditional fire management processes. It also serves as an educational tool for traditional fire managers by providing an alternative or complementary practice to prescribed burning.[137] Following the 2018 Nowra workshops, over 90% of the participants said that the event increased their knowledge of indigenous fire techniques and over 60% said they are likely to change their own fire management practices as a result of the workshop.[138]

## Challenges and Future Research

In order to scale cultural burning practices like firestick farming, more institutional support is necessary. This support should come in the form of firm and measurable policy directives that have the ability to provide resources and build capacity for the effort. As of now, the cultural burning initiatives across Australia are decentralized and operate fairly independent of one another.[139]

Furthermore, firestick farming can be a time-intensive and expensive activity, as many small fires are involved; this differs from traditional fire management activities which typically involve larger, singular events. As a result, only a small fraction of the land currently undergoes cultural burning, and many of these sites do not receive multi-year funding for continued maintenance.[140]

Lastly, there is still resistance to using burning practices like firestick farming for wildfire management due to general fears about the risk of fire. In many places, especially in densely settled urban areas with overlapping uses and interests, the use of fire to fight fire is often considered a risky endeavor. Thus, it has been difficult to expand the practice to areas unfamiliar with indigenous land management work.

1 mi

Klamath River Valley

*Fire Lighting*

Yurok Territory

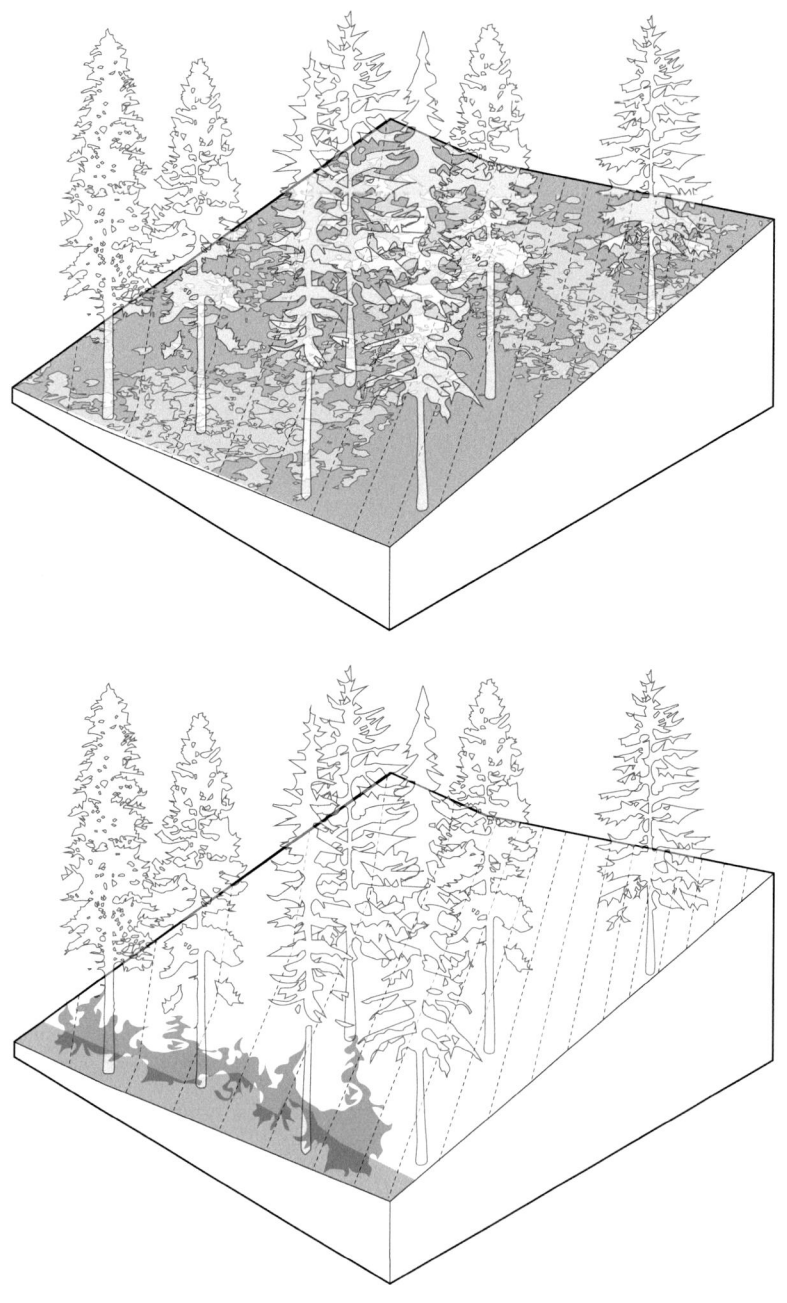

21

*Fire Lighting*

| | |
|---|---|
| Description | Prescribed Fire Training Exchanges and cooperative burns involving a wide range of stakeholders |
| Example Location | Yurok Territory, California |
| Size | 10 acres |
| Primary Implementer | TREX program |
| Stakeholders and Team Members | Cultural Fire Management Council, Yurok Tribal members, CALFire, The Nature Conservancy, Air Quality Management District, USFS Ranger District, landowners, USDA Forest Service, Department of the Interior, private fire professionals |

## Technique Overview

Prescribed Fire Training Exchanges (TREX) is a unique fire management model that brings together a wide array of stakeholders – including tribal members, private landowners, representatives from federal and state agencies, and private fire professionals – to conduct burns together. The TREX initiative emerged out of the Fire Learning Network (FLN), which has members from TNC, the USFS, and various agencies under the U.S. Department of the Interior (DOI).[141]

During TREX events, participants share knowledge with one another in order to support the use of fire on the landscape for a range of ecological, social, and economic needs. Workshops are designed to be collaborative learning environments where participants collectively build capacity. Participants are intentionally grouped into diverse squads, with each squad representing a range of stakeholders and a range of experience levels.

## Planning and Design Process

Figure 5.28 (*Previous*) Aerial of Yurok Territory. Source: Google Earth 2022.

Figure 5.29 (*Left*) A view of a densely forested landscape (top) and the same view when using hybrid burning practices to reduce fuel loads while also supporting indigenous cultural traditions (bottom).

In the spring of 2021, a TREX event took place on a ten-acre burn site located on the eastern side of the Klamath River Valley along California State Highway 169, and situated within the Yurok territory. The western edge of the unit abuts the highway and the Yurok wildland fire station sits in the middle of the unit, with multiple onsite water sources. It is a remote site in a densely forested landscape with steep terrain and a southwesterly aspect. The canopy primarily consists of Douglas fir trees (Pseudotsuga menziesii) with an understory primarily composed of Himalayan blackberry (Rubus armeniacus), and hazel (Corylus sp). Due to the large presence of hazel and the tribal history with the site, the land is of high cultural value for the Yurok community and benefits from the presence of fire.[142]

For this particular burn, the TREX team outlined two general goals and four specific objectives. The two resource goals were to use fire to support the growth of hazel, a culturally-significant species for the Yurok tribe, and use fire to reduce <u>fuel</u> in the forest that could contribute to larger and more intense wildfires. The specific objectives were to: (1) reduce dead plant material <u>fuel</u> on the ground by 30–90%, (2) reduce low competing vegetation 20–80% six months after the burn, (3) maintain at least 80% of hardwood species one year after the burn, and (4) maintain at least 80% of legacy trees one year after the burn.[143]

Prior to the burn, a <u>handline</u> was built around the sides of the unit to create a safety zone around the area and the site was routinely monitored to ensure that the environmental conditions met the prescription criteria outlined in the <u>burn plan</u>. Then, on the day of the burn, participants primarily used <u>hand firing</u> via drip torches to ignite the unit. The first technique that was employed is called <u>backing</u> – with this technique, igniters burned in a line against the dominant wind direction. This is considered a cool, slow burn that removes deadwood and litter from the forest floor. This technique is often employed first to create a wide safe zone at the edge of the unit. For this particular site, the backing lines started at the top of the slope. After establishing a safe zone through <u>backing</u>, the team employed what is called a <u>head fire</u> which burns with the wind, moving faster and hotter. Lastly, to connect the back fire to the head fire, the team employed a <u>flanking fire</u> technique, running burns perpendicular to the wind direction.[144]

During the exercise, participants called holders ensured proper containment by paying attention to sharp corners, high <u>fuel</u> areas, and areas with legacy trees, structures, or infrastructure. Participants also monitored fuel moisture, weather patterns, <u>fire behavior</u>, and smoke dispersal. Once the burn was complete, the team conducted a mop-up of the site and patrolled for <u>hotspots</u>, rehabilitated the <u>fireline</u> around the periphery, felled any damaged trees, monitored participants, and documented the process in a report.[145]

Currently, the Yurok community has over 20 tribal members that have received training via the TREX initiative and there is a growing desire to continue the collaborative burning efforts.

## Performance and Evaluation

The TREX program is unique in that it functions as a bridge to connect people interested in the practice of burning and to have them learn from one another. It creates a more diverse group of practitioners, fosters intergenerational learning, increases knowledge sharing and adaptive capacity, and expands networking opportunities for those involved. Based on a recent survey of TREX participants, 85% of respondents were very satisfied with the experience and 99.5% would recommend the program to a friend.[146]

The TREX program is also a highly flexible initiative that promotes on-the-ground learning through doing. Furthermore, the strategies can adjust and adapt with each burn, with participants learning as they go.

TREX also helps to preserve the rich cultural traditions of the Yurok community. Not only is this technique focused on <u>fuel reduction</u>, it is also focused on supporting the continuance of indigenous life.[147] According to recent studies, burning practices through programs like TREX can increase the growth of high-quality stems sought after for basket framing, up to 13-fold in just one year after a burn.[148] Furthermore, burns can aid in supporting traditional food sources for the Yurok tribe by managing pests and supporting rich foraging grounds and can also be a valuable employment opportunity for tribal members interested in burning practices and forestry.

Burning programs like TREX also improve the ecological function of the landscape in a number of different ways. First, by creating a patchy mosaic in the forest, burns can support ecological complexity and increase biodiversity through the creation of improved wildlife habitat.[149] They can optimize soil productivity, decrease the potential for erosion, and increase the amount of water going to legacy trees and into nearby waterways. They can help to preserve large trees that serve to sequester carbon in the soil and they can also help support fire-dependent plant species. Lastly, these kinds of burns can actually improve long-term air quality.[150]

TREX burns help to reduce the potential of large, intense wildfires by creating fuel breaks in the forest. This technique also has the potential to increase public safety and to reduce the risk to firefighters who are often on the front line during wildfires.[151]

## Challenges and Future Research

One challenge for TREX relates to the professional workforce associated with burns. Given the education and training needed for professional certification, and the risks associated with the work, it can be difficult to recruit and expand the workforce.[152]

Another barrier relates to funding. Currently, most TREX events are funded by securing grants or through donations and gifts.

Additionally, when compared to other techniques, burning initiatives like TREX require a significant amount of planning before the burn can take place. Team members must develop an in-depth burn plan and seek approval from a range of local, regional, and state agencies. It is also difficult to get insurance for burning events like TREX, so liability is a challenge.

Safety concerns are another challenge for programs like TREX. Team members are often working in remote, steep, and highly vegetated terrain. Also, while not very common, burn escapes can jeopardize nearby landscapes, structures, and infrastructure; furthermore, they can profoundly impact people's perception about the risks associated with burns. One final safety consideration relates to air quality and smoke generated from burns.[153]

## Fire Lighting Interview

Figure 5.30 Margo Robbins, Co-Founder and President of the Cultural Fire Management Council (CFMC), Yurok Tribal Member.

DESIGN BY FIRE: Can you talk about the importance of fire in your community?

MARGO ROBBINS: I live in the Upper Yurok reservation, which is a very remote, mountainous area. I am a basket maker, and basket making is a very important part of our culture. We make a basket when a baby is born, and we also use them for storage, for trapping animals, and as prayer items. The frames of the baskets are made with hazel which needs to be burned in order to send up new roots. In addition to weavers needing materials, we were also afraid that our elders might not be able to escape if there was a wildfire. A small group of us were just determined that we will continue to put fire on the land so that basket weavers will have weaving materials, so that our traditional food sources and medicine plants will be available, and so much more. And as we came down this path, we came to realize that it's more than just that – that restoring the land with fire is actually an act of restoring the people.

DF: Can you talk a little bit about what a TREX burn looks like?

MR: Only qualified firefighters are allowed to participate in TREX. They come from all over the United States, and sometimes abroad, and we typically will accept around 30 people per burn. The TREX model means that people come to get trained to increase their qualifications.

During our welcome talk we make sure that everyone understands the reason that they're here is not just to burn brush, but that our culture relies on putting fire on the land. At lunch, we'll have one of the ceremonial dance leaders talk to them about how dance is connected to culture and to fire – that way they really make that connection between what they're doing and our cultural lifeways. After that we'll go out and walk along the <u>fireline</u> where we intend to burn. Depending on the area, we'll point out culturally important species found there or explain that we need to clear the area for bigger prairies, for example, so that elk will return to the area. In addition to cultural ties, they'll also talk about firing patterns, hazards to be aware of, and weather patterns. In the evening, we'll go back to camp for dinner and cultural presentations. I'll often do a basket presentation, or we'll have a story-teller or another activity like cracking and grinding acorns which we can cook for everyone. We mix cultural information with ecosystem restoration information, along with how to do a burn in a safe way.

The second day, we have a suite of training sessions and groups will go from one to the next. We have training on the fire engine, the condensed fire safety booklet, radio communications, and what to do if someone gets hurt. We also build on a sense of teamwork – everyone has to know everybody else's name, because when you're out there you need to know who you're responsible for. On the third day, we have an early breakfast, everyone is given their assignments, and we discuss the <u>burn plan</u>. Then we go out to the site and do a quick-run as well as another briefing. Then we start putting fire down, starting at the top of a slope and placing parallel lines coming downward. At the end of the day, we circle up and we talk about what went well and how we can improve. My colleague Rick always says "Everytime we have a TREX, it's the best one we've ever done!" So we're always improving.

DF: What other benefits can you share in terms of using this cooperative model?

MR: The really huge benefit is that when people come here to burn, even after 20+ years of experience, they learn about the benefits of fire for the first time. They have to really overcome this life-long teaching that fire is bad, and that it must be attacked. One of the key concepts in the fire safety booklet is "Fight fire aggressively," but here we're actually there to *light* fire and work with it. It's a very different mindset.

The other part is this: for over 100 years, fire has been held tightly in the hands of government agencies. We're teaching them that there are other people and other entities who have the necessary expertise, and that needs to be acknowledged. In fact, fire needs to be returned to the hands of the people. Our family burn program is a great example of that. Why shouldn't people be able to learn how to use fire responsibly and use it to restore or maintain their land and protect their homes? There's so much land the government can't ever have enough time or resources to take care of it the way it needs to be taken care of, they need to share that responsibility with people who know how to help.

Some other benefits include water and climate change. When these giant wildfires are burning up an entire forest, not only are they putting all of these toxins and smoke into the air but they're also burning up the trees that would normally be sequestering carbon. Burning can help prevent some of this. Also when we're burning at a landscape level it leaves charcoal on the ground, so we are actually purifying the water on a landscape level.

Finally, there's the benefit of community employment. My goal was to train up 20 local people so that we can do our own burns, and we did meet that goal a few years ago. Now with a grant from CAL FIRE we're able to hire on five full-time workers, which is our Fire and Fuel crew, as well as a Fire Coordinator, Fire and Fuels manager, and myself as the Executive Director as part-time employees, as well as a number of people that we hire on just when we're burning.

1 mi

Espolla

*Fire Flocking*

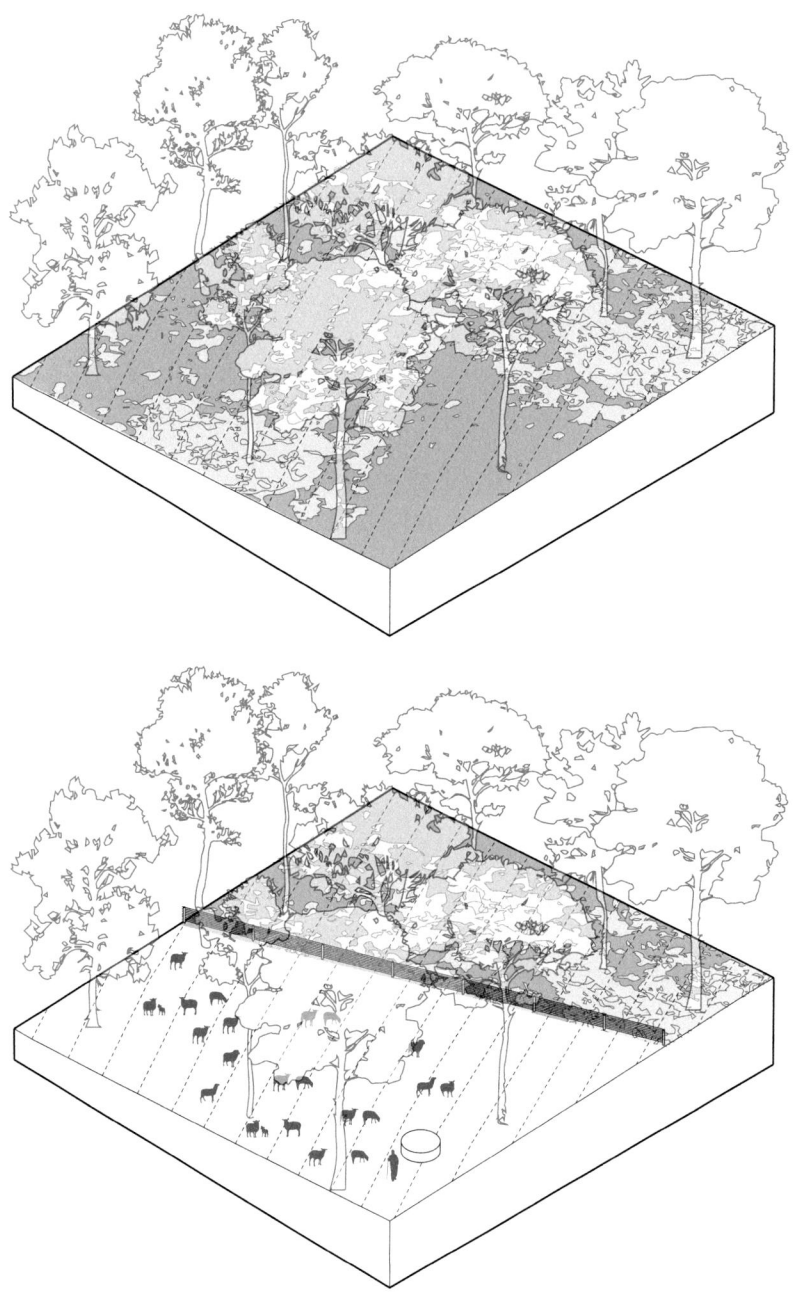

22

*Fire Flocking*

| | |
|---|---|
| Description | Grazing of animals in overgrown forests and creating a bio-based economy |
| Example Location | Catalonia Region, Spain |
| Size | 5,000 acres |
| Primary Implementer | The Pau Costa Foundation |
| Stakeholders and Team Members | Private landowners, wildfire management services, firefighters, the government of Catalonia, local butchers, local shepherds (with their sheep, goats, and cows), and local restaurants and shops |

## Technique Overview

In an effort to create a more fire-resistant landscape in the region of Catalonia, Pau Costa Foundation is leading a project called Fire Flocks (Ramats de Foc).[154] The initiative started in 2016, with four years of funding from the Fundación Daniel y Nina Carasso, and has five primary goals: (1) manage overgrown vegetation in former agricultural areas to reduce the amount of fuel in the landscape; (2) bring back the culture and function of silvopastoralism - practices of integrating vegetation management and the grazing of domesticated animals in mutually beneficial ways - by reintroducing cattle, goats, and sheep; (3) increase biodiversity by creating strategic openings in the forest cover and a mosaic-like pattern across the landscape; (4) create a unique brand of value-added agricultural products from the effort; and (5) educate the public about the project, the landscape changes facing the region, and how they can personally contribute.[155]

## Planning and Design Process

To initiate zones for the project, local firefighters consult the team on strategic areas that could benefit from fuel reduction via grazing. These areas, called strategic management points (SMP), are critical zones that could change the behavior of wildfire or help with firefighter staging and operation efforts during a wildfire event.[156] Once identified, these zones first undergo a mechanical fuel reduction process to allow for grazing to happen. Then, local shepherds are identified. Ideally, these shepherds own land in the SMP areas, but if this is not the case, the shepherds can sign a five-year lease with private landowners to use their land for grazing.

One goal of the project is to create vertical and horizontal fuelbreaks in the forest. To do this, shepherds work with Pau Costa foundation to design a grazing plan that details grazing sub-areas, numbers of animals, the rotational schedule, feed supplements, and infrastructure needs.[157] Throughout this process, the shepherds are given latitude to use their own expertise to conduct the management as they see fit. Before animals are brought

Figure 5.31 (*Previous*) Aerial of Catalonia. Source: Google Earth 2022.

Figure 5.32 (*Left*) A view of an overgrown forest (top) and the same view when using grazing animals like cows, sheep, and goats to reduce fuel and support a bio-based economy (bottom).

to the area, infrastructure such as paving, water points, and fencing are planned and developed to ensure a smooth operation. After a year of grazing, an inspection certifies a 90% reduction in annually grown herbaceous material and a 60% reduction in annually grown shrub material on the landscape. This inspection typically happens prior to the summer months when wildfires are common. If the inspection is successful, shepherds typically receive around $40 per acre.

Shepherds involved in the project are allowed to use the official brand, which links agricultural products to wildfire management efforts in the region. Then, based on the product type, herbivore breed, and flock size, shepherds prepare their goods over the course of the year and target butchers, shops, and restaurants appointed by Pau Costa Foundation.

Since its initial pilot, the project has expanded: as of 2021, the project has recruited 22 shepherds with cattle, goats, and sheep, and over 60 local restaurants, butchers, and shops. With this experience, Pau Costa Foundation is also involved in a larger network of grazing organizations across Europe that contributes to knowledge-building. For example, the Fire Shepherds program seeks to educate European shepherds beyond traditional milk and meat production and toward management practices that promote biodiversity and wildfire risk reduction. In turn, the shepherds can diversify the services they offer to their clients and spread the idea of shepherding for multiple functions. The Fire Shepherds program is also creating a model for shepherding schools based on best practices to further this vision.

## Performance and Evaluation

Studies have shown that <u>fuel reduction</u> activities like extensive grazing can slow the spread of wildfire and can decrease the intensity of wildfire.[158] Furthermore, by reducing the <u>fuel</u> continuity both horizontally and vertically, the risk of <u>crown fires</u> is reduced.[159] Also, by developing breaks in the forest, the project has also created new habitat with a thinner understory.

The project has also uplifted an agricultural tradition in the region of Catalonia by helping to rejuvenate a rural bioeconomy. By boosting shepherding and associated products, the project has supported local food production chains. In turn, shepherds in the program have received access to private property for grazing, exposure to new services and value-added products, and new landscape management knowledge for bolstering wildfire adaptation.[160]

The technique has also been considered fairly low-cost, especially after the initial <u>mechanical thinning</u> phase, and scalable to other fire-prone areas of Europe. Furthermore, the agricultural products branded with the project label have seen an increase in demand. For example, following the pilot project, there was an increase of 12% in meat sales and a 40% increase in the number of restaurants selling the products.[161]

Lastly, the program has helped to educate people in the region about wildfire management and has invited them to get involved. The team has found innovative ways to raise awareness about wildfire including the development of product labels that inform end consumers of the project and how they are contributing to landscape management with their purchase.[162]

## Challenges and Future Research

One challenge for the project is that many of the stakeholders hold different perspectives and interests; extensive private ownership in rural areas also makes it difficult for the team to manage the landscape. For example, it is common for 20–30 landowners to have property in one SMP. Furthermore, it can be difficult to find shepherds willing to partake in the program given that traditional shepherding is a time-consuming job that requires a lot of experience and is not very profitable.[163] Creating more of a financial incentive for shepherds could help with this hurdle. Furthermore, many of the SMPs identified for the project are difficult to access and require infrastructural planning to allow for grazing to happen.

While the herbivores selected for the program are already adapted to the climate and topography of Catalonian region, there is a range of productivity and efficiency in the flocks.[164] Part of this variability stems from palatability as herbivores prefer different kinds of plants. For example, sheep tend to eat grass, shoots, and other small understory plants whereas goats tend to eat brush and broadleaf vegetation. Another reason for this range in productivity is associated with access and terrain. For example, due to their size, cattle may have difficulty accessing certain areas of the SMPs due to steep slopes and natural obstacles.[165]

While grazing is considered a cost-effective method for reducing fuel and the products associated with the program are increasing in demand, an in-depth cost-efficiency analysis might be helpful to assess financially-sustainable models for implementing Fire Flocking and how to develop pathways to get there.[166]

Another challenge with the program stems from the annual evaluation process to determine the effectiveness of the grazing regimes on the landscape. The current evaluation model uses simplified indicators that do not fully address the complexity of the management process. The timing of the annual inspections is also problematic now that the Mediterranean fire season is no longer confined to summer months. Additionally, due to the late start of the pilot in 2016 and a lack of historical data of grazing in the SMPs,[167] it is difficult to make assessments of the long-term wildfire adaptation performance on the landscape.

Finding the right venues to sell the products is challenging. There is also a need to refine the target audience for the products beyond the typical urban, middle-aged consumers. Additionally, there is a desire to develop an everyday product that is more present on the consumer side and can appeal to a wide range of shops and restaurants.

The last challenge relates to scalability. Moving forward, it might be helpful to better understand how SMP sizes and distribution across the landscape relates to wildfire management at a regional scale.

## Fire Flocking Interview

Figure 5.33 Marta Carola, Biologist and Shepherd.

DESIGN BY FIRE: Can you tell me a little bit about your role and what project you're working on?

MARTA CAROLA: I work in the province of Girona, in Catalonia, Spain. There is a local cow called the Albera that's in danger of extinction, so we created a breeding project to save this cow. In addition to the breeding project, we're also trying to restore the land. In the last century, many farmers left for the cities and their fields were abandoned, so now the forest has invaded the property and there's a lot of dense growth of bushes and trees. By using the cows to graze the underbrush, we're trying to create more fields to provide food for the cows as well as trying to stop potential fires from spreading.

DF: How many animals are you working with on the property?

MC: Right now we have 500 cows, plus the calves, on 5,000 acres.

DF: How do you manage the grazing of those animals? Are they on a rotational schedule?

MC: That's our goal. We're clearing many acres to create fields for the cows, and we're also building fences to make a rotation.

Before beginning all this work, I read about how big an optimal grazing space should be, and how many spaces I would need in order to make an effective rotation. So with that, I made a map with all the spaces we should clear. At the moment, I use

Google maps to show all the areas that I'm working in to clear the brush before grazing, and we mark the areas that we've already created the fence for.

DF: I'm curious as to how you might measure how wildfire has been reduced in the region as a result of the work you're doing?

MC: The firemen actually marked on some of our maps where the most dangerous areas are where fire could begin. We were already bringing our cows to eat there, but now we clear these areas even more with a machine and we force the cows graze there for a few days to totally clear the area. I don't know if it will stop a fire, but at least we will help to slow it down. Before this, the cows were a bit more free to graze where they want but now that we are collaborating with the Fire flocking project, we are trying to work more consciously.

DF: What can you say about the community response to this program? When you go to sell your products, do people recognize the Fire flocks brand or is it new to people still?

MC: That's what the foundation is working on. I think it's still quite new, the idea that animals are clearing the forest in order to stop the fires. I think there is still a lot of work to do in order to make people more aware.

I can say that more organizations, including the government, are working in that direction. Because the fires now are so big, and we have a very dense forest and strong winds here that makes it very difficult to stop the fires. That's why more people are talking about putting animals in the forest, to work on fire prevention rather than attacking the fire.

DF: Are there any other challenges that you can speak to in regards to your work?

MC: It's difficult to continue with this many animals, so I'm stuck between saving the breeding project and having enough food for the animals here. But I'd say the main challenges are getting enough money to hire people to work with me, and to find the people who want to do this work.

It's very difficult to find good shepherds. For many years, being a shepherd was not desirable work. It's hard – you're in the mountains all day, it's wild, and you don't get good pay. So little by little, this work has disappeared. But the good thing is that now there is a shepherding school here, so more young people who want to do this work can be trained.

DF: I'm wondering if there is any talk about collaboration with the government to encourage this kind of work? Maybe providing incentives to do this work or other types of funding?

MC: I know many people agree that the government should encourage the owners of the forests to clear them and maintain them to prevent fires. But as of now there's not a lot of help or incentives for these owners. If you have a forest and you can't earn any money from it, you will abandon it. So I've read about the idea of the government giving incentives to the landowners to do this work, because it's a way of helping the whole community.

# Notes

1   Philip Gibbons et al., "Options for Reducing House-Losses during Wildfires without Clearing Trees and Shrubs," *Landscape and Urban Planning* 174, (2018): 10–17.
2   Ibid.
3   Ibid.
4   Ibid.
5   Ibid.
6   Ibid.
7   Ibid.
8   Ibid.
9   Ibid.
10  Ibid.
11  Conservation Management, and NSW National Parks and Wildlife Service, "Native Shelterbelts: Benefits for Wildlife," accessed October 13, 2021, https://www.environment.nsw.gov.au/resources/nature/Factsheet7NativeShelterbelts.pdf
12  Hayley Johnson et al., "Landcare Notes: Shelterbelt Design," accessed October 13, 2021, https://crec.ifas.ufl.edu/extension/windbreaks/pdf/Victoria%20Dept%20Sustain&Environ-Shelterbelt%20Design%20LC0136.pdf
13  Stephen Murphy, "Farm Plantations Can Reduce Bushfire Risk," accessed October 13, 2021, https://www.recreatingthecountry.com.au/blog/farm-plantations-can-reduce-bushfire-risk
14  Stephen Murphy, "Deciduous Trees Can Provide Crucial Bushfire Protection in Rural Australia," accessed October 13, 2021, https://www.recreatingthecountry.com.au/blog/deciduous-trees-can-provide-crucial-bushfire-protection
15  Murphy, "Farm Plantations."
16  Murphy, "Deciduous Trees."
17  Johnson, "Landcare Notes."
18  Dean Turner, "Towards a More Fire Retardant Landscape with Shelterbelting," accessed October 13, 2021, https://www.thecrossingland.org.au/wp-content/uploads/2021/04/Towards-a-more-Fire-Retardant-Landscape-with-Shelterbelts-1.pdf
19  Conservation Management, "Native Shelterbelts."
20  Ibid.
21  Ibid.
22  Ibid.
23  Ibid.
24  Johnson, "Landcare Notes."
25  Conservation Management, "Native Shelterbelts."
26  Johnson, "Landcare Notes."
27  Conservation Management, "Native Shelterbelts."
28  Ibid.
29  Johnson, "Landcare Notes."
30  Deep Green Permaculture, "Australian Native."
31  Conservation Management, "Native Shelterbelts."
32  John Mason, "Using Australian Plants for a Fire Break," accessed October 13, 2021, https://prwire.com.au/pr/40105/using-australian-plants-for-a-fire-break
33  Conservation Management, "Native Shelterbelts."
34  Mason, "Using Australian Plants."
35  Ibid.
36  Conservation Management, "Native Shelterbelts."
37  Dean Turner, interview with the author, July 25, 2021.
38  "Vegetation Management Best Practices for Transmission System Operators," LIFE Elia-RTE, accessed October 12, 2021, http://www.life-elia.eu/en/Vade-mecum-Best-practices-for-Transmission-System-Operators
39  "Ponds and Invasive Species under High Voltage Lines," LIFE Elia-RTE, accessed October 12, 2021, http://www.life-elia.eu/en/Brochure-n-5-Ponds-and-invasive-species
40  LIFE Elia-RTE, "Vegetation Management."
41  Ibid.
42  "Layman's Report," LIFE Elia-RTE, accessed October 12, 2021, http://www.life-elia.eu/_dbfiles/lacentrale_files/1400/1424/LIFE%20Elia-RTE_Layman%20report%202018_simple_EN_HD.pdf
43  Ibid.
44  Final Event of the Life Elia-RTE Project in Drôme, accessed February 14, 2021, http://www.life-elia.eu/en/News/Final-event-of-the-LIFE-Elia-RTE-project-in-Drome
45  LIFE Elia-RTE, "Ponds and Invasives."
46  Ibid.
47  Ibid.
48  Ibid.
49  LIFE Elia-RTE, "Layman's Report."
50  Ibid.

51 LIFE Elia-RTE, "Ponds and Invasives."
52 Ibid.
53 LIFE Elia-RTE, "Layman's Report."
54 LIFE Elia-RTE, "Ponds and Invasives."
55 LIFE Elia-RTE, "Layman's Report."
56 LIFE Elia-RTE, "Vegetation Management."
57 Ibid.
58 "Cleland National Park Management Plan 2022," accessed November 1, 2022, https://cdn.environment.sa.gov.au/environment/docs/Cleland-National-Park-management-plan-2022.pdf
59 Jordan Duke, "The Digital & The Wild: Mitigating Wildfire Risk Through Landscape Adaptations," accessed November 1, 2022, https://www.asla.org/2016studentawards/186884.html
60 "Cleland National Park."
61 Duke, "The Digital."
62 Ibid.
63 Emily Davis et al., "Collaboration and Stewardship Authority: The Ashland Forest Resiliency Project," accessed October 22, 2021, https://scholarsbank.uoregon.edu/xmlui/bitstream/handle/1794/19626/FS_9_AFR.pdf?sequence=1&isAllowed=y
64 Kerry Metlen et al., "Integrating Forest Restoration, Adaptation, and Proactive Fire Management: Rogue River Basin Case Study," *Canadian Journal of Forest Research* 51, (2021): 1292–1306.
65 Kerry Metlen and Darren Borgias. *Ashland Forest Resiliency Stewardship Project Monitoring Plan* (Ashland: Ashland Forest Resiliency Stewardship Project, 2013).
66 Marty Main, *Block 3 Silvicultural Prescriptions – Draft* (Ashland: Ashland Forest Resiliency Stewardship Project, 2012).
67 Ashland Forest Resiliency Stewardship Project, *Project Factsheet: AFR Project Multiparty Monitoring* (Ashland: Ashland Forest Resiliency Stewardship Project, 2021).
68 Metlen et al., *Ashland Forest.*
69 Ashland, *Project Factsheet.*
70 Ibid.
71 Ashland Forest Resiliency Stewardship Project, *Ashland Forest All Lands Restoration* (Ashland: Ashland Forest Resiliency Stewardship Project, 2017).
72 Kerry Metlen, interview with author, September 2, 2021.
73 Ashland, *Project Factsheet.*
74 Davis et al., "Collaboration and Stewardship."
75 Ashland, *Ashland Forest.*
76 Metlen et al., "Integrating Forest."
77 Kerry Metlen, interview with author, September 2, 2021.
78 Metlen et al., "Integrating Forest."
79 Kerry Metlen, interview with author, September 2, 2021.
80 Metlen et al., "Integrating Forest."
81 Ibid.
82 Kerry Metlen, interview with author, September 2, 2021.
83 Ibid.
84 Patricia Holmes et al., "Setting Restoration Priorities for the Cape Floristic Region, using Cape Town as an Example" (Clarens, SA: Presentation, Biodiversity Planning Forum, May 8, 2013).
85 Table Mountain National Park, *Tokai and Cecilia Management Framework 2005–2025* (Cape Town: South African National Park, 2009).
86 Table Mountain, *Tokai and Cecilia* (Ibid.).
87 Ibid.
88 Philippa Huntly, "Clearing Table Mountain," *Veld & Flora*, March 2005, 28–30.
89 Alanna Rebelo, interview with author, August 20, 2021.
90 Galloway et al., "The Impact."
91 Huntly, "Clearing."
92 "FOTP Chairman's Report 2019," Friends of Tokai Park, accessed October 12, 2021, https://tokaipark.com/wp-content/uploads/2021/03/Chairpersons-Report-FoTP-AGM-2019.pdf
93 Table Mountain, *Tokai and Cecilia.*
94 Friends of Tokai Park, "FOTP Chairman's Report."
95 Table Mountain, *Tokai and Cecilia.*
96 Friends of Tokai Park, "FOTP Chairman's Report."
97 "Chairs Report. Friends of Tokai Park," Friends of Tokai Park, accessed October 12, 2021, https://tokaipark.com/wp-content/uploads/2021/03/chairs-report-FoTP-2020.pdf
98 Alanna Rebelo, interview with author, August 20, 2021.
99 Table Mountain, *Tokai and Cecilia.*
100 Friends of Tokai Park, "Chairs Report."
101 Louise Stafford, *Invasive Alien Species Strategy for the Greater Cape Floristic Region* (Cape Town: Western Cape Nature Conservation Board t/a CapeNature, 2009).
102 City of Cape Town, *City of Cape Town.*
103 Stafford, *Invasive.*

104 Table Mountain, *Tokai and Cecilia.*

105 Alanna Rebelo, interview with author, August 20, 2021.

106 Arjan Meddens et al., "Fire Refugia: What are They, and Why Do They Matter for Global Change?" *BioScience* 68, no.12 (2018): 944–954.

107 "Bush Regeneration Handbook: A Guide for Local Volunteer Groups," accessed November 1, 2022, https://www.landcareillawarra.org.au/wp-content/uploads/BushRegenerationManual.pdf

108 T.C. Fuller and G. Douglas Barbe, "The Bradley Method of Eliminating Exotic Plants from Natural Reserves," accessed November 1, 2022, https://s3.wp.wsu.edu/uploads/sites/2062/2014/04/bradleytechnique.pdf?x96359#:~:text=The%20Bradley%20method%20makes%20practical,the%20regeneration%20of%20native%20plants

109 "Bush Regeneration Handbook."

110 Fuller and Barbe, "The Bradley Method."

111 Meddens et al., "Fire Refugia."

112 "Bush Regeneration Handbook."

113 Ibid.

114 Simon Pooley, "Recovering the Lost History of Fire in South Africa's Fynbos," *Environmental History* 17, (January 2012): 55–83.

115 Craig M. Mulqueeny et al., "Landscape-Level Differences in Fire Regime between Block and Patch-Mosaic Burning Strategies in Mkuzi Game Reserve, South Africa," *African Journal of Range & Forage Science* 27, no.3 (2010): 143–150.

116 Pooley, "Recovering."

117 Stuart Hall et al., "A Dynamic Modeling Tool to Anticipate the Effectiveness of Invasive Plant Control and Restoration Recovery Trajectories in South African Fynbos," *Restoration Ecology* 29, no.3 (2020): 1–13.

118 Carly Cowell et al., "A Ranking System for Prescribed Burn Prioritization in Table Mountain National Park, South Africa," *Journal of Environmental Management* 190, (2017): 283–289.

119 Mulqueeny, "Landscape-Level."

120 Cowell, "A Ranking."

121 Ibid.

122 Ibid.

123 Mulqueeny, "Landscape-Level."

124 Pooley, "Recovering."

125 Cowell, "A Ranking."

126 Ibid.

127 Pooley, "Recovering."

128 Aaron Petty, "Introduction to Fire-Stick Farming," *Fire Ecology* 8, no.3, (2012): 1–2.

129 *National Indigenous Fire Workshop*, Firesticks Alliance, accessed May 23, 2022, https://firesticks.app.box.com/s/okl4pc5v9x6yd9q2cymx1ses4acbwi0u

130 Lorena Allam, "Right Fire for Right Future: How Cultural Burning can Protect Australia from Catastrophic Blazes," *The Guardian*, accessed May 23, 2022, https://www.theguardian.com/australia-news/2020/jan/19/right-fire-for-right-future-how-cultural-burning-can-protect-australia-from-catastrophic-blazes

131 Will Smith et al., "Persuasion without Policies: The work of Reviving Indigenous Peoples' Fire Management in Southern Australia," *Geoforum* 120, (2021): 82–92.

132 Firesticks Alliance, *National Indigenous.*

133 Ibid.

134 Smith, "Persuasion."

135 R. Bliege Bird et al., "The 'Fire Stick Farming' Hypothesis: Australian Aboriginal Foraging Strategies, Biodiversity, and Anthropogenic Fire Mosaics," *PNAS* 105, no.39 (2008):14796–14801.

136 Ibid.

137 Smith, "Persuasion."

138 Firesticks Alliance, *National Indigenous.*

139 Smith, "Persuasion."

140 Ibid.

141 Andrew G. Spencer et al., "Enhancing Adaptive Capacity for Restoring Fire-Dependent Ecosystems: The Fire Learning Network's Prescribed Fire Training Exchanges," *Ecology and Society* 20, no.3 (2015): 1–13.

142 Phillip Dye, *Prescribed Fire Burn Plan: Hwy 169* (2020).

143 Ibid.

144 Ibid.

145 Ibid.

146 Andrew G. Spencer, "Building Capacity and Integrating Training, Education and Experience: The Fire Learning Network's Prescribed Burn Training Exchanges," Master's Thesis, (Colorado State University, 2014).

147 Margo Robbins, "Learning Networks: Notes from the Field," accessed October 12, 2021, https://www.conservationgateway.org/ConservationPractices/FireLandscapes/FireLearningNetwork/USFLNPublications/Documents/161-NotesFromTheField-YurokTREX-Fall2019.pdf

148 Marks-Block et al., "Revitalized Karuk and Yurok Cultural Burning to Enhance California Hazelnut for Basketweaving in Northwestern California, USA," *Fire Ecology* 17, no.6 (2021):1-20.
149 Spencer, "Building Capacity."
150 Dye, *Prescribed Fire.*
151 Ibid.
152 Spencer, "Building Capacity."
153 Black et al., "Organizational Learning from Prescribed Fire Escapes: A Review of Developments Over the Last 10 Years in the USA and Australia," *Current Forestry Reports* 6, (2020): 41–59.
154 "Ramats de Foc," accessed October 12, 2021, https://www.ramatsdefoc.org/en/project/
155 Frank Krumm, Andreas Schuck and Andreas Rigling, *How to Balance Forestry and Biodiversity Conservation – A View Across Europe* (Birmensdorf: European Forest Institute and Swiss Federal Research Institute WSL, 2020).
156 Mario Colonico et al., "The Role of Fuel Management Smart Solutions in Mitigating Fire Risk: A Review" (COST Action CA18135 "FIRElinks" Short Term Scientific Mission, 2018).
157 Krumm et al., *How to Balance.*
158 Krumm et al., *How to Balance.*
159 Colonico et al., "The Role of Fuel."
160 Krumm et al., *How to Balance.*
161 Ibid.
162 Krumm et al., *How to Balance*, and "La Fonda Gràfica," accessed October 12, 2021, https://lafondagrafica.com/en/portfolio/herds-of-fire/
163 Colonico et al., "The Role of Fuel."
164 Aura Secanell Perarnau, "Ramaderia per a La Prevenció D'incendis Forestals: Revisió bibliogràfica," accessed October 12, 2021, https://ddd.uab.cat/pub/tfg/2013/123713/TFG_asecanellperarnau.pdf
165 Ibid.
166 Colonico et al., "The Role of Fuel."
167 Krumm et al., *How to Balance.*

# Bibliography

*A Guide to Integrated Fire Management.* Cape Town: FynbosFire, 2016.

Allam, Lorena. "Right Fire for Right Future: How Cultural Burning can Protect Australia from Catastrophic Blazes," *The Guardian.* Accessed May 23, 2022. https://www.theguardian.com/australia-news/2020/jan/19/right-fire-for-right-future-how-cultural-burning-can-protect-australia-from-catastrophic-blazes

Ashland Forest Resiliency Stewardship Project. *Ashland Forest All Lands Restoration.* Ashland: Ashland Forest Resiliency Stewardship Project, 2017.

Ashland Forest Resiliency Stewardship Project. *Project Factsheet: AFR Project Multiparty Monitoring.* Ashland: Ashland Forest Resiliency Stewardship Project, 2021.

Bird, R. Bliege, Douglas Bird, Brian Codding, Charles Parker and James Jones. "The "Fire Stick Farming" Hypothesis: Australian Aboriginal Foraging Strategies, Biodiversity, and Anthropogenic Fire Mosaics." *PNAS* 105, no.39 (2008): 14796–14801.

Black, Anne, Peter Hayes, and Roger Strickland. "Organizational Learning from Prescribed Fire Escapes: A Review of Developments Over the Last 10 Years in the USA and Australia." *Current Forestry Reports* 6, no.1 (2020): 41–59.

"Bush Regeneration Handbook: A Guide for Local Volunteer Groups." Accessed November 1, 2022. https://www.landcareillawarra.org.au/wp-content/uploads/BushRegenerationManual.pdf

"Cleland National Park Management Plan 2022." Accessed November 1, 2022. https://cdn.environment.sa.gov.au/environment/docs/Cleland-National-Park-management-plan-2022.pdf

Colonico, Mario, Marta Serra Davos, and Eduard Plana Bach. "The Role of Fuel Management Smart Solutions in Mitigating Fire Risk: A Review." COST Action CA18135 "FIRElinks" Short Term Scientific Mission, 2018.

Conservation Management, and NSW National Parks and Wildlife Service. "Native Shelterbelts: Benefits for Wildlife." Accessed October 13, 2021. https://www.environment.nsw.gov.au/resources/nature/Factsheet7NativeShelterbelts.pdf

Cowell, Carly and Chad Cheney. "A Ranking System for Prescribed Burn Prioritization in Table Mountain National Park, South Africa." *Journal of Environmental Management* 190 (2017): 283–289.

Davis, Emily Jane, and Eric White. "Collaboration and Stewardship Authority: The Ashland Forest Resiliency Project." Accessed October 22, 2021. https://scholarsbank.uoregon.edu/xmlui/bitstream/handle/1794/19626/FS_9_AFR.pdf?sequence=1&isAllowed=y

Deep Green Permaculture. "Australian Native and Exotic Fire Resistant Trees and Plants for Fireproof." Accessed October 13, 2021. https://deepgreenpermaculture.com/2020/02/25/australian-native-and-exotic-fire-resistant-trees-and-plants-for-fireproof-landscapes/

Duke, Jordan. "The Digital & The Wild: Mitigating Wildfire Risk Through Landscape Adaptations." Accessed November 1, 2022. https://www.asla.org/2016studentawards/186884.html

Dye, Phillip. *Prescribed Fire Burn Plan: Hwy 169* (2020).

Final Event of the Life Elia-RTE Project in Drôme. Accessed February 14, 2021. http://www.life-elia.eu/en/News/Final-event-of-the-LIFE-Elia-RTE-project-in-Drome

Friends of Tokai Park. "Chairs Report. Friends of Tokai Park." Accessed October 12. 2021. https://tokaipark.com/wp-content/uploads/2021/03/chairs-report-FoTP-2020.pdf

Friends of Tokai Park. "FOTP Chairman's Report 2019." Accessed October 12, 2021. https://tokaipark.com/wp-content/uploads/2021/03/Chairpersons-Report-FoTP-AGM-2019.pdf

Fuller, T.C. and G. Douglas Barbe, "The Bradley Method of Eliminating Exotic Plants from Natural Reserves." Accessed November 1, 2022. https://s3.wp.wsu.edu/uploads/sites/2062/2014/04/bradleytechnique.pdf?x-96359#:~:text=The%20Bradley%20method%20makes%20practical,the%20regeneration%20of%20native%20plants

Galloway, Alistair, Patricia Holmes, Mirijam Gaertner, and Karen Esler. "The Impact of Pine Plantations on Fynbos above-Ground Vegetation and Soil Seed Bank Composition." *South African Journal of Botany* 113 (2017): 300–307.

Gibbons, Philip, Malcolm Gill, Nicholas Shore, Max Moritz, Stephen Dovers, and Geoffrey Cary. "Options for Reducing House-Losses During Wildfires Without Clearing Trees and Shrubs." *Landscape and Urban Planning* 174 (2018): 10–17.

Hall, Stuart, Rita Bastos, Joana Vicente, Ana Sofia Vaz, João P. Honrado, Patricia M. Holmes, Mirijam Gaertner, et al. "A Dynamic Modeling Tool to Anticipate the Effectiveness of Invasive Plant Control and Restoration Recovery Trajectories in South African Fynbos." *Restoration Ecology* 29, no.3 (2020): 1–13.

"Herds of Fire: Fire-Fighting Products Sold Only at Butchers." Accessed October 12, 2021. http://gremicarn.com/ramats-de-foc-productes-contra-incendis-venuts-nomes-a-les-carnisseries/

Holmes, Patricia and Anthony Rebelo. "Setting Restoration Priorities for the Cape Floristic Region, Using Cape Town as an Example." Clarens, SA: Presentation at the Biodiversity Planning Forum, May 8, 2013.

Huntly, Philippa. "Clearing Table Mountain." *Veld & Flora*, March 2005: 28–30.

Johnson, Hayley, and James R. Brandle. "Landcare Notes: Shelterbelt Design." Accessed October 13, 2021. https://crec.ifas.ufl.edu/extension/windbreaks/pdf/Victoria%20Dept%20Sustain&Environ-Shelterbelt%20Design%20LC0136.pdf

Krumm, Frank, Andreas Schuck, and Andreas Rigling. How to Balance Forestry and Biodiversity Conservation – A View Across Europe. Birmensdorf: European Forest Institute and Swiss Federal Research Institute WSL, 2020.

"La Fonda Gràfica." Accessed July 12, 2021. https://lafondagrafica.com/en/portfolio/herds-of-fire/

LIFE Elia-RTE. "Layman's Report: Creation of Green Corridors for Biodiversity under High-Voltage Lines." Accessed July 12, 2021. http://www.life-elia.eu/en/Layman-report

LIFE Elia-RTE. "Vegetation Management Best Practices for Transmission System Operators." Accessed July 12, 2021. http://www.life-elia.eu/en/Vade-mecum-Best-practices-for-Transmission-System-Operators

LIFE Elia-RTE. "Ponds and Invasive Species Under High Voltage Lines." Accessed July 12, 2021. http://www.life-elia.eu/en/Brochure-n-5-Ponds-and-invasive-species

Main, Marty. *Block 3 Silvicultural Prescriptions – Draft*. Ashland: Ashland Forest Resiliency Stewardship Project, 2012.

Marks-Block, Tony, Frank K. Lake, Rebecca Bliege Bird, and Lisa M. Curran. "Revitalized Karuk and Yurok Cultural Burning to Enhance California Hazelnut for Basketweaving in Northwestern California, USA." *Fire Ecology* 17, no.6 (2021): 1–20.

Mason, John. "Using Australian Plants for a Fire Break." Accessed October 13, 2021. https://prwire.com.au/pr/40105/using-australian-plants-for-a-fire-break

Meddens, Arjan, Crystal Kolden, James Lutz, Alistair Smith, Alina Cansler, John Abatzoglou, Garrett Meigs, et al. "Fire Refugia: What are they, and Why do They Matter for Global Change?" *BioScience* 68, no.12 (2018): 944–954.

Metlen, Kerry. Interview with the author, September 2, 2021.

Metlen, Kerry and Darren Borgias. *Ashland Forest Resiliency Stewardship Project Monitoring Plan*. Ashland: Ashland Forest Resiliency Stewardship Project, 2013.

Metlen, Kerry L., Terry Fairbanks, Max Bennett, Jena Volpe, Bill Kuhn, Matthew P. Thompson, Jim Thrailkill et al. "Integrating Forest Restoration, Adaptation, and Proactive Fire Management: Rogue River Basin case study." *Canadian Journal of Forest Research* 51 (2021): 1292–1306.

Mulqueeny, Craig M., Paul S. Goodman, and Timothy G. O'Connor. "Landscape-Level Differences in Fire Regime between Block and Patch-Mosaic Burning Strategies in Mkuzi Game Reserve, South Africa." *African Journal of Range & Forage Science* 27, no.3 (2010): 143–150.

Murphy, Stephen. "Deciduous Trees Can Provide Crucial Bushfire Protection in Rural Australia." Accessed October 13, 2021. https://www.recreatingthecountry.com.au/blog/deciduous-trees-can-provide-crucial-bushfire-protection

Murphy, Stephen. "Farm Plantations Can Reduce Bushfire Risk." Accessed October 13, 2021. https://www.recreatingthecountry.com.au/blog/farm-plantations-can-reduce-bushfire-risk

*National Indigenous Fire Workshop*, Firesticks Alliance. Accessed May 23, 2022. https://firesticks.app.box.com/s/okl4pc5v9x6yd9q2cymx1ses4acbwi0u

Perarnau, Aura Secanell. "Ramaderia per a La Prevenció D'incendis Forestals: Revisió bibliogràfica." Accessed October 12, 2021. https://ddd.uab.cat/pub/tfg/2013/123713/TFG_asecanellperarnau.pdf

Petty, Aaron. "Introduction to Fire-Stick Farming." *Fire Ecology* 8, no.3 (2012): 1–2.

Pooley, Simon. "Recovering the Lost History of Fire in South Africa's Fynbos." *Environmental History* 17 (January 2012): 55–83.

"Ramats de Foc." Accessed October 12, 2021. https://www.ramatsdefoc.org/en/project/

Rebelo, Alanna. Interview with the Author. August 20, 2021.

Robbins, Margo. "Learning Networks: Notes from the Field." Accessed October 12, 2021. https://www.conservationgateway.org/ConservationPractices/FireLandscapes/FireLearningNetwork/USFLNPublications/Documents/161-NotesFromTheField-YurokTREX-Fall2019.pdf

Robbins, Margo. Interview with the author, July 17, 2021.

Smith, Will, Timothy Neale, and Jessica Weir. "Persuasion without Policies: The Work of Reviving Indigenous Peoples' Fire Management in Southern Australia." *Geoforum* 120 (2021): 82–92.

Spencer, Andrew G. "Building Capacity and Integrating Training, Education and Experience: The Fire Learning Network's Prescribed Burn Training Exchanges." Master's Thesis, Colorado State University, 2014.

Spencer, Andrew G., Courtney A. Schultz, and Chad M. Hoffman. "Enhancing Adaptive Capacity for Restoring Fire-Dependent Ecosystems: The Fire Learning Network's Prescribed Fire Training Exchanges." *Ecology and Society* 20, no.3 (2015): 1–13.

Stafford, Louise. *Invasive Alien Species Strategy for the Greater Cape Floristic Region.* Cape Town: Western Cape Nature Conservation Board t/a CapeNature, 2009.

Steinberg, Michele. "Firewise Forever? Voluntary Community Participation and Retention in Firewise Programs." *Proceedings of the Second Conference on the Human Dimensions of Wildland Fire* (2011): 79–87.

Table Mountain National Park. *Tokai and Cecilia Management Framework 2005–2025.* Cape Town, South African: National Park, 2009.

Turner, Dean. Interview with the author, July 25, 2021.

Turner, Dean. "Towards a More Fire Retardant Landscape with Shelterbelting." Accessed October 13, 2021. https://www.thecrossingland.org.au/wp-content/uploads/2021/04/Towards-a-more-Fire-Retardant-Landscape-with-Shelterbelts-1.pdf

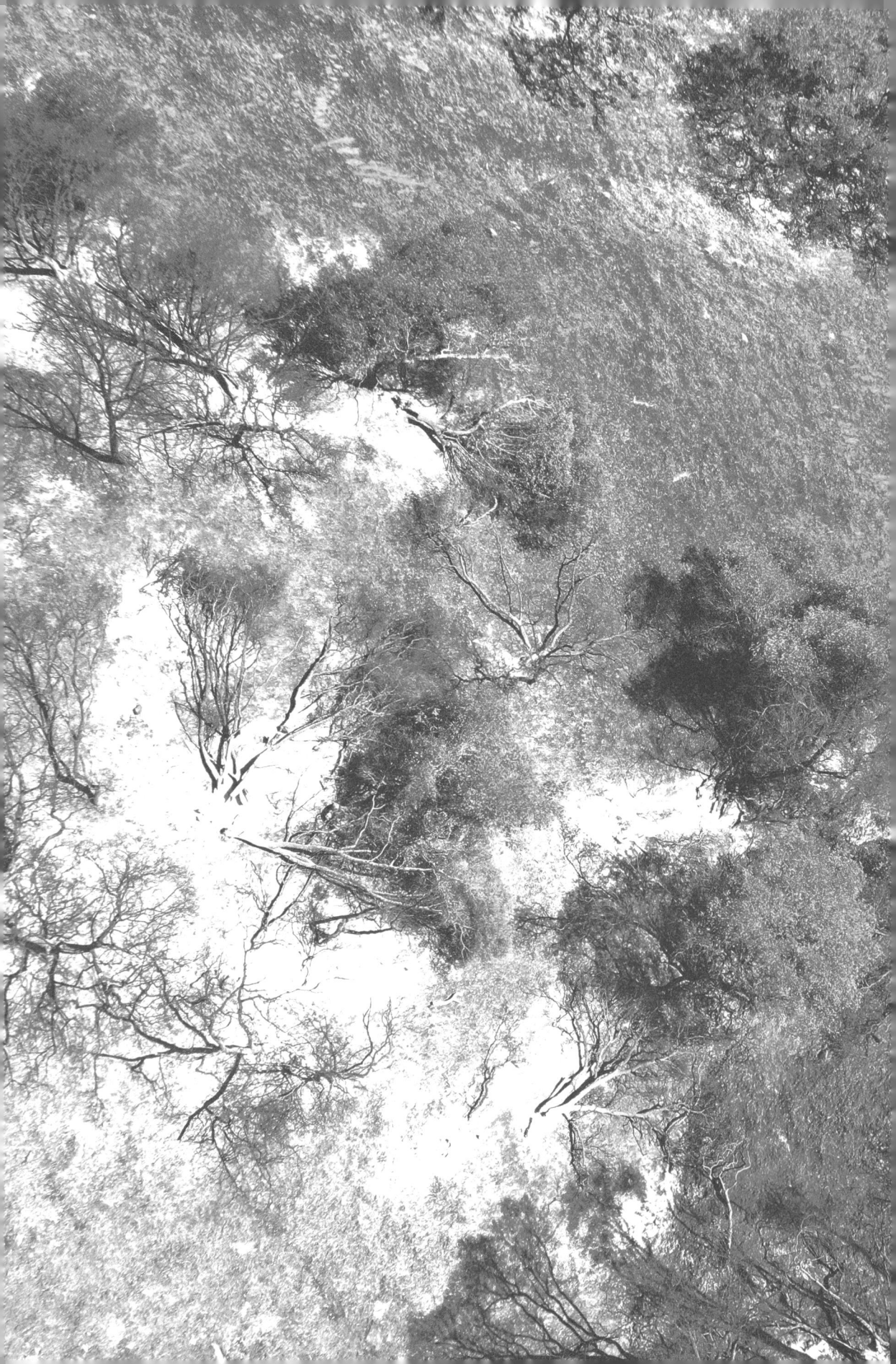

Retreat

Figure 6.1
View of a landscape around Lake Berryessa after the LNU Lightning Complex Fires of 2020.

Figure 6.2
View of a remaining foundation near Lake Berryessa following the LNU Lightning Complex Fires of 2020.

Figure 6.3
View of a playground structure following the Ferguson Fire of 2018.
Photograph by Hanna Prissen.

Figure 6.4
Feral wildlands in the Sierra Nevada mountain range that are allowed to burn.
Photograph by Derek Young.

Figure 6.5
View of a remaining pool near Lake Berryessa following the LNU Lightning Complex Fires of 2020.

Figure 6.6
View of a valley regularly burned by fire.
Unidentified photographer.

# Chapter 6

# Retreat

*an act of moving back or withdrawing, often from something dangerous, difficult, or unsustainable; the process of receding from a position or state attained.*

Retreat can be defined as the intentional, coordinated movement of people, buildings, communities, and infrastructures away from areas and landscapes of perceived high risk. It is most commonly understood and discussed in relation to flood risk and sea level rise, but is increasingly applicable across a wide and increasing range of environmental risks, such as drought, excessive heat, toxicity, and wildfire. In terms of fire, retreat can also mean withdrawing from actively managing landscapes that formerly were. In contrast to resistance approaches, retreat is an` intentional "letting go" of perceived control or dominion of a landscape; of giving it over to itself to evolve and become. In most contexts, retreat is the very opposite of maintaining the cultural and ecological status quo.

DOI: 10.4324/9781003172956-9

1 mi

Terrasanta

*Development Limiting*

Mission Trails Regional Park

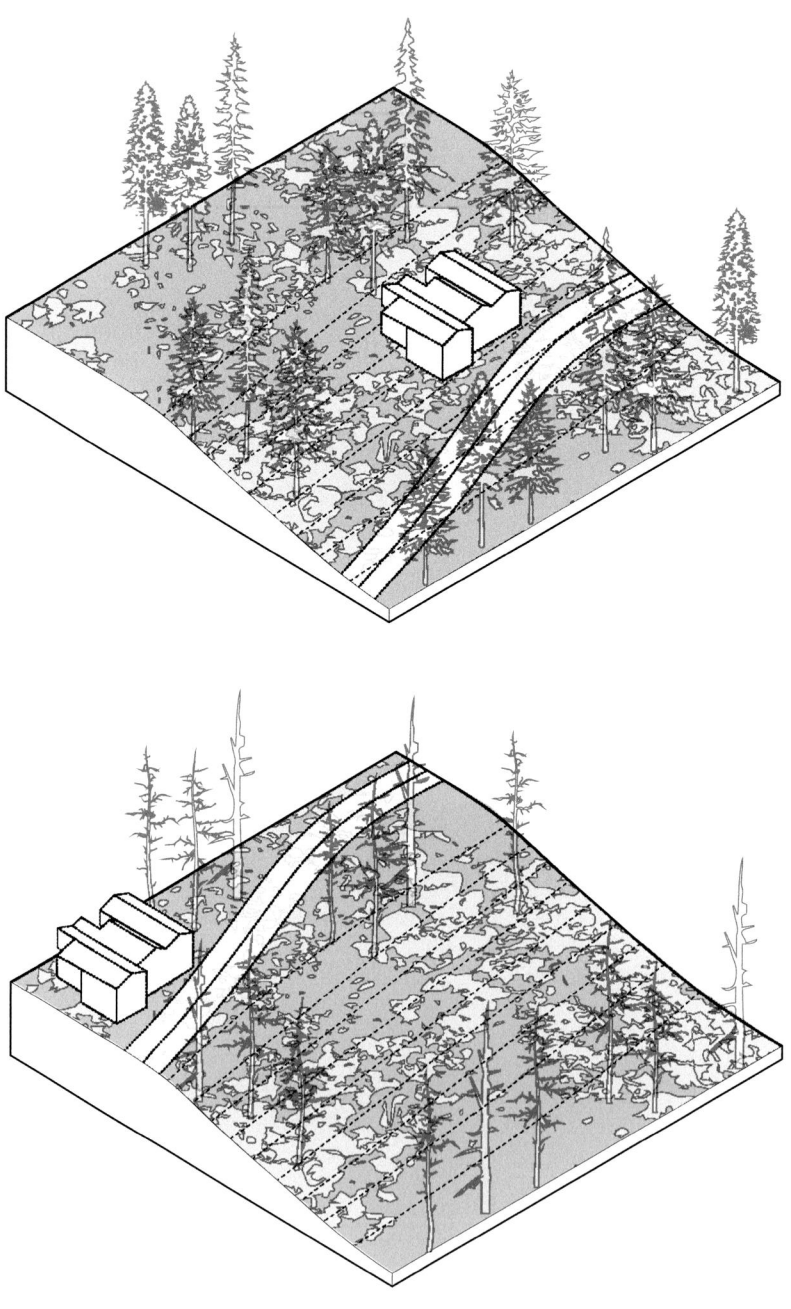

*Development Limiting*

| | |
|---|---|
| Description | Creating stricter development standards for new construction |
| Example Location | San Diego, California |
| Size | Varies |
| Primary Implementer | City of San Diego |
| Stakeholders and Team Members | Developers, private landowners |

## Technique Overview

New development in fire-prone areas has the potential to increase ignition risk and expose more people to wildfire. One technique used to curb this trend is the creation of stricter development standards for new construction. By integrating knowledge of fire spread into land use planning and regulation tools, communities can protect their residents from dangerous development patterns.

These stricter standards can take many forms but typically involve changing permissible site selection – restricting new construction in areas of higher wildfire risk and instead, requiring the clustering of new construction in areas of lower wildfire risk. Steep slope guidelines are one example of this.

In 1997, the City of San Diego adopted guidelines for new development on the city's steep hillsides. Under these guidelines, any project proposed on or near significantly sloped land must conform to certain development standards in order to receive a construction permit.[1]

## Planning and Design Process

Steep slopes are a concern for new construction because topography is one of the three main elements of the fire triangle. Along with weather and fuel, topography can play a key role in the initial ignition, the intensity of a burn, the direction and speed of an event, and the management of the wildfire. This is especially true in wildfires that are not wind-driven, as these events tend to follow topography and burn upslope. Generally, wildfires burn faster and more intensely as the steepness of a slope increases. This condition creates longer flame lengths that reach upslope, preheating and igniting vegetative fuels – both surface and canopy – as well as structural fuels. Thus, any building situated on a steep slope has a higher risk of damage from a wildfire.[2]

Some steep slope guidelines allow for mid-slope construction, but stipulate stricter building material standards and vegetation maintenance standards. For example, since steeper slopes are harder to defend, it is often suggested that plants and trees downslope of a structure should have more horizontal spacing between them. This suggested spacing increases with a rise in slope. Some guidelines also suggest that mid-slope homes should go through hardening measures.

For the City of San Diego, the steep hillside guidelines restrict construction in areas with significant grade change. The guidelines define steep hillsides as sites with a gradient of at least 25% and a vertical elevation of at least 50 feet. The guidelines also apply if part of a site has a natural slope of at least 200% and a vertical elevation of at least ten feet. In most cases, new buildings cannot encroach into these zones and must be located at least 30 feet from the hillside.[3]

Steep hillside guidelines, like those used in the City of San Diego, guide the layout of planned unit developments (PUDs), large residential subdivisions with shared amenities. With these kinds of developments, the guidelines help determine the location and layout of lots and associated roads to reduce public safety concerns.[4]

## Performance and Evaluation

Development-limiting regulations like steep hillside guidelines are successful in reducing some risk for new construction, especially in areas that are particularly susceptible to wild-fire events. Preventing new construction on steep slopes also makes it easier for wildfire professionals to manage fires and protect community assets.

Beyond wildfire risk reduction, these regulations have other community benefits. Environmentally, they can prevent excessive cut and fill practices that often come with sloped developments; these extensive grading practices can negatively affect existing drainage patterns, increase sedimentation, reduce water quality, and make the landscape more susceptible to post-fire flooding, landslides, and debris flows. Development-limiting guidelines can also protect special vegetation and animal communities by preserving environmentally sensitive habitat found on undeveloped land.[5]

In many cases, development-limiting regulations like steep hillside guidelines can reduce construction costs and simplify implementation. For instance, it is generally easier and more affordable to extend water and sewer infrastructure across flat land than across sloped terrain.[6]

Lastly, there are cultural benefits related to these types of regulations. Steep hillside guidelines, for example, preserve the quality of existing views. They also can protect important recreational opportunities found in sloped areas.[7]

## Challenges and Future Research

While development-limiting policies like steep hillside guidelines prevent new construction in some of the most hazardous conditions, they often cannot eliminate risk. Furthermore, guidelines like these may promote development in potentially unsafe areas. For example, if a house is setback from the top of a steep slope, this buffer may not be able to stop an intense wildfire. Or if a house sits on a flat ridge with appropriate setbacks around it, it may still have exposure to wildfire on all sides.[8] And even if a house sits on the gentle leeward side of a steep slope, it may still have significant wildfire risk due to wind turbulence at the top of the crest.

These kinds of policies can also provide a false sense of security for developers and homeowners who think that proper site selection will eliminate wildfire risk. In reality, none of these techniques can stand by themselves. Instead, those wishing to build new construction should consider a range of risk-reduction measures like proper site selection, the use of <u>fire-resistant</u> building materials, and the maintenance of <u>defensible space</u>.

Another drawback to regulations like this is that they can limit community growth opportunities. This is especially a challenge in places like California that urgently need more housing to offset current shortages.

1 mi          Bonita

*Construction Halting*

Site of Otay Ranch Resort Village

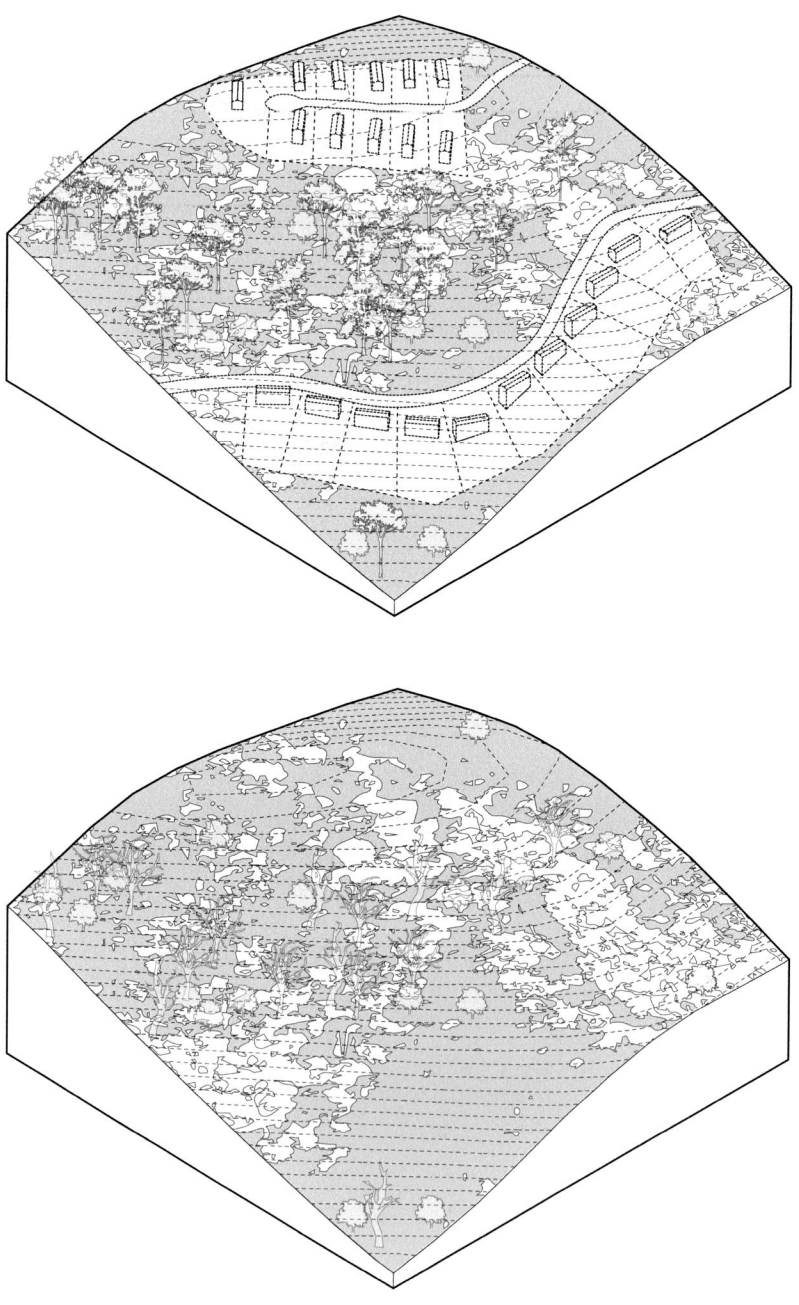

# 24

## Construction Halting

| | |
|---|---|
| Description | Stopping proposed developments for failing to address wildfire risk |
| Example Location | Chula Vista, California |
| Size | 1869 acres |
| Primary Implementer | Chula Vista |
| Stakeholders and Team Members | Developers, private landowners |

## Technique Overview

Despite increasing wildfire risk, development in the wildland-urban interface (WUI) continues to grow across the American West. In recent years, courts have stopped many high profile projects in California because their Environmental Impact Reports (EIRs) fail to adequately address the increased wildfire risk that the projects create.

One example of this is a proposed project called the Otay Ranch Resort Village located in Proctor Valley in Southwestern San Diego County, east of Chula Vista, and immediately south of Jamul. The project is part of a large, 23,000 acre development plan that the Board of Supervisors and City Council jointly adopted in 1993. The larger plan includes a number of urban villages, specialty villages, estate parks, business parks, and commercial centers.[9]

Otay Ranch Resort Village, one of the proposed urban villages, covers about 8% of the larger development area and is located on a relatively flat mesa extending southward from the Jamul Mountains. The mesa has several steep canyons that drain into the Lower Otay Reservoir, located just south and west of the site. Coastal sage scrub and grassland affected by grazing cover a significant percentage of the site.[10] The village plan proposed 1,119 residential units, 10,000 square feet of commercial space geared toward neighborhood use, a fire and police station, an elementary school, 24 acres of parks, 776 acres of open space, and 1,284 acres of undisturbed landscape.[11]

The site currently sits in a very high fire hazard severity zone that has a long history of wildfires – there are records of 68 fire perimeters within five miles of the project site. These fires include two large fires that burned a majority of the proposed project area – the Mine/Otay Fire in 2003, and the Harris Fire in 2007.[12] The existing fire risk for this particular site is so high because of its highly flammable vegetation but also because of its topography. The steep slopes that cut through the mesa can function as fire corridors that disperse embers far ahead of the fire front, spreading fire faster through the landscape.

## Planning and Design Process

In December 2020, a month after the San Diego County Board of Supervisors approved the Otay Ranch Resort Village EIR, a group of environmental groups – Endangered Habitats

Figure 6.9 (*Previous*) Aerial of San Diego. Source: Google Earth 2022.

Figure 6.10 (*Left*) A view of a neighborhood being planned in a highly vulnerable location (top) and the same view with the development no longer being proposed in the fire-prone landscape (bottom).

League, California Native Plant Society, Center for Biological Diversity, Preserve Wild Santee, California Chaparral Institute, and the Sierra Club – filed a lawsuit claiming that the EIR did not properly address the project's impacts on the landscape.[13] Three months later, the California Attorney General's Office, who oversees the California Environmental Quality Act (CEQA), filed a motion to join the lawsuit.

In October 2021, the lawsuit made it to the San Diego Superior Court. Here, the judge ruled in favor of the plaintiffs and struck down the San Diego County Board of Supervisor's approval of the development. In the ruling, he addressed a number of concerns. He agreed that the project EIR did not acknowledge how the project, itself, would increase wildfire risk and that it did not provide analysis of existing wildfire risk on the property prior to the proposed mitigation efforts. In his ruling, he also cited insufficient greenhouse gas mitigation plans, the disturbance of sensitive habitat, an issue of cumulative project impacts, incompatibility with the County's General Plan, and an inadequate provision of affordable housing.[14]

## Performance and Evaluation

Turning to the court system to halt proposed development projects with increased wildfire risks helps to prevent future losses of life and property. Not only are many of these projects, like Otay Ranch Resort Village, located in existing fire hazard severity zones, they also put more people in harm's way and increase the likelihood of a human-ignited event. This technique also has the potential to reduce the cost that taxpayers might need to pay to help maintain and protect risky developments in the case of a wildfire event.

The rulings of these cases can set a precedent for future litigation action in the WUI. Projects similar to Otay Ranch Resort Village that courts have recently halted in California include Guenoc Valley Project in Lake County, Fanita Ranch Project in Santee, Valiano Project in San Diego, and Harmony Grove Village South in San Diego.

Beyond wildfire risk reduction, this construction halting technique can also help to protect existing sensitive habitat, and reduce the extensive environmental impacts related to sprawl development.

## Challenges and Future Research

Critics of construction halting measures often claim that the techniques exacerbate the housing crisis. This is argued in places like California, where a lack of housing has pushed the state to reduce local growth controls. It's claimed that projects like Otay Ranch Resort Village provide much needed housing, even if just by providing "move-up" housing for those wishing to upgrade – opening up more affordable housing elsewhere in the region. Critics also claim that this kind of legal action in the WUI might deter other developers from pursuing projects in the region and that halting development projects could reduce future employment opportunities for residents in the region, especially if the project includes schools, commercial areas, industrial areas, or resort areas.

Beyond litigation, risky developments could be halted in other ways. For example, cities and counties could use basic urban planning tools like general plans, zoning, and subdivision ordinances to limit new sprawl in <u>fire hazard severity zones</u>. Fire-prone landscape zoning acts, similar to past legislation that has helped limit new development along coasts, on wetlands and along earthquake faults, could be developed. Urban growth boundaries could also be adopted locally or mandated by states. Furthermore, incentives for local governments to adopt these frameworks could be provided through planning and technical assistance grants or preference for competitive infrastructure funding. At the same time, states or federal agencies could refuse to make such funding available for local authorities that enable development in severe-risk areas.

1 mi

Paradise

*Incentivized Relocating*

Concow

25

*Incentivized Relocating*

| | |
|---|---|
| Description | Encouraging people to move out of wildfire's way and relocate |
| Example Location | Paradise, California |
| Size | Variable |
| Primary Implementer | Paradise Recreation and Parks District |
| Stakeholders and Team Members | Private landowners, The Nature Conservancy and Conservation Biology Institute |

## Technique Overview

Following catastrophic floods across the U.S., local governments have worked with FEMA to offer eligible homeowners the pre-disaster value of their home in exchange for not rebuilding. To date, this type of federal-backed voluntary buyout program has yet to be implemented for wildfire areas to help people move out of harm's way, but some vulnerable communities have developed their own. The community of Paradise is one example.

On November 8, 2018, an electrical spark ignited ten miles northeast of Paradise. In less than six hours, it swept up the ravine to the ridge and destroyed 90% of the town. In total, the Camp Fire took 85 lives and destroyed more than 19,000 structures.[15]

On the day that the fire erupted, Governor Newsom declared a State of Emergency in Butte County. Four days later President Trump made a Major Disaster Declaration for the State of California.[16] FEMA then initiated its Long-Term Community Recovery Program to help with the recovery and rebuilding process. For six months, community leaders met with residents to create a collective vision and to prioritize next steps in a long-term recovery plan, which was adopted by the Paradise Town Council in June 2019.[17]

The plan outlines five primary goals – the top one being to "Make Paradise Safer," especially in the face of future wildfire events. In this section, it discusses a range of techniques including preventative measures like landscape fuel management projects to reduce risk.[18]

Figure 6.11 (*Previous*) Aerial of Paradise. Source: Google Earth 2022.

Figure 6.12 (*Left*) A view of a cul-de-sac prior to a wildfire event (top) and the same view after a wildfire event when residents were encouraged to move out of harm's way and relocate (bottom).

## Planning and Design Process

One year following the adoption of the long-term recovery plan, the Paradise Recreation and Parks District put forth a wildfire risk reduction proposal to buy up properties at the edge of town and turn this land into a large defensible space around the community. According to the proposal, this space could serve as a community firebreak, or a wildfire risk reduction buffer (WRRBs), to help slow or stop a fire front, and as a staging area for fire professionals to safely maintain events. Furthermore, the space could become active recreational or agricultural land and could serve as an urban growth boundary to prevent development from creeping into adjacent wildlands.[19]

The process for developing the WRRB started with prioritizing parcels at the edge of Paradise for wildfire risk reduction. To do this, the team developed a wildfire probability model and began mapping parcels that were next to and downwind of high-risk areas. They then compared a no-action scenario to six <u>fuel</u> management scenarios to understand fire risk reduction potential.[20]

After identifying parcels, they then reached out to property owners to gauge interest in selling their land. If the property owners were willing, they purchased the land using grant money and donations. A year into the program, they had acquired 300 acres of land on the periphery of town.[21]

## Performance and Evaluation

While the purchasing of privately owned parcels at the edge of a community and converting this land into defensible space has yet to be systematically reviewed in scientific literature, a significant amount of gray literature supports the development of wildfire buffers for reducing wildfire risk. This space can also serve as <u>community refuge</u> spaces during wildfire events and staging areas for professional firefighting crews,[22] And other potential co-benefits include increasing open space for residents, limiting sprawl, reducing development-related edge effects, and conserving habitat.[23]

Furthermore, while no federal-backed <u>voluntary buyout program</u> exists for wildfire-impacted areas, researchers are studying the feasibility of this kind of program in places like Paradise – indicating a potential shift in policy.

## Challenges and Future Research

Incentivized relocating programs like the one being implemented in Paradise face a number of challenges. First, they are primarily reactive and may only be successful after a catastrophic wildfire event – thus, they might be difficult to implement in any sort of preventative way.

Also, if the purchased parcels are converted into a <u>defensible space</u> buffer, this belt may provide a false sense of security for remaining residents. Even if the buffer is a mile wide, it cannot eliminate the threat from <u>ember attacks</u>, which can launch embers miles ahead of the <u>fire front</u>. It also may not be able to protect the community in the case of extreme future conditions.

Furthermore, this scale of a project can be difficult logistically. In Paradise, over 30,000 acres were identified for the WRRB, but a year into the project, only 1% of the land area had been acquired. Furthermore, almost 95% of the land identified is privately-owned, making parcel acquisition a challenging process.[24] Additionally, given that there is no federal support, the community of Paradise may need more money to be able to buy up land. Lastly, the population of Paradise is not even 30% of what it once was before the fire, making it one of the first documented cases of voluntary retreat. Given that a significant people are leaving, does an intervention at this scale even make sense, especially with its high maintenance costs?

Beyond voluntary buyout programs, there are other ways to incentivize residents to move away from wildfire-prone areas. Removing government-backed fire insurance plans or instituting variable fire insurance rates based on risk could also encourage people to avoid high-risk areas. Another potential tool is a "transferable development rights" framework. Under such a framework, developers wishing to build more intensively in lower-risk town centers could purchase development rights from landowners in rural areas where fire-prone land is to be preserved or returned to unbuilt status. The rural landowners are thus compensated for the lost use of their property. These frameworks have been used in Montgomery County, Maryland, and in Massachusetts and Colorado.

1 mi

Lake Felicity

Isle de Jean Charles

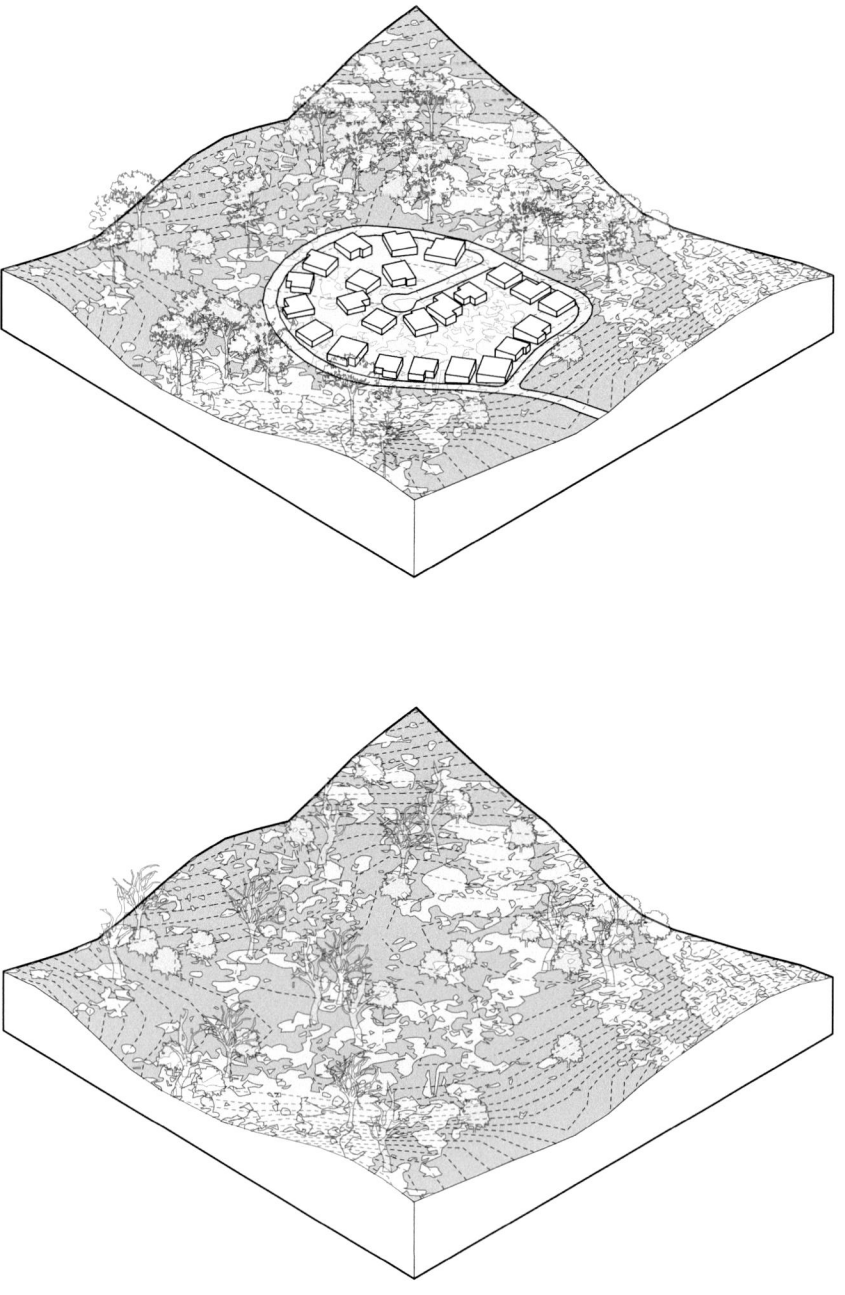

## *Wholesale Moving*

| | |
|---|---|
| Description | Managing the entire resettlement of a vulnerable community |
| Example Location | Isle de Jean Charles, Louisiana |
| Size | 320 acres |
| Primary Implementer | State of Louisiana |
| Stakeholders and Team Members | Biloxi-Chitimacha-Choctaw tribe, CSRS, The Lowlander Center, Evans + Lighter, Waggonner & Ball, OLIN, APTIM, HR&A, SGD, CDBG, KyleWhyte, GeoEngineers |

## Technique Overview

Vulnerable communities may want to relocate but do not want to leave neighbors and friends. This technique involves managing the entire resettlement of a vulnerable community.

While this technique has yet to be implemented for wildfire-prone areas, there is a long history of its use after catastrophic floods. One place it is currently being used is Isle de Jean Charles, Louisiana where residents are members of the Biloxi-Chitimacha-Choctaw tribe. This tribe has had a presence on the isle for over 200 years, earning a livelihood primarily through fishing and trapping.

Since 1955, the island has lost 98% of its landmass due to coastal erosion, storm inundation, and sea level rise. During extreme weather events, flooding often blocks vehicular access to the island, stranding residents and cutting them off from resources like schools and hospitals.[25]

With this increased risk of flooding, many residents have voluntarily moved away from the island, but some cannot afford to leave or are reluctant to move because of their attachment to the landscape and their community. While there have been two efforts to resettle the community over the last two decades, they were unsuccessful due to a lack of a full buy-in from the tribe and resistance from nearby communities about the location of the new settlement.[26] Currently, a third attempt is underway and is the U.S.' first federally funded, community-led resettlement project driven by climate change.[27]

Figure 6.13 (*Previous*) Aerial of Isle de Jean Charles. Source: Google Earth 2022.

Figure 6.14 (*Left*) A view of a community prior to a wildfire event (top) and the same view after a wildfire event when all residents were resettled in a new, less vulnerable location (bottom).

## Planning and Design Process

In 2016, the state of Louisiana won a $92 million grant through the National Disaster Resilience Competition to bolster their preparations for future flood-related disasters. The larger project – called the Louisiana Strategic Adaptations for Future Environments Program (LA SAFE) – focuses on three goals: to reshape, to retrofit, and to resettle. Areas identified for reshaping are those that have less than three feet of flooding during a 100-year storm event 50 years from now. These areas are reserved for increased development and economic growth. Areas identified for retrofitting are those that have between 3 and 14 feet of flooding. These areas are targeted for restructuring, either through protective means like levees or policy means like land use planning. Areas identified for resettling are those that might suffer from over 14 feet of flooding. For these areas, moving residents to less vulnerable locations is the strategy.[28]

The authors of LA SAFE identified Isle de Jean Charles as a prime location to test the third goal of resettling, and earmarked $48 million for this process. The project began in 2016 with data gathering and engagement. Then, from 2016–2019, the team selected a site for resettlement, acquired it, and began the master planning process. In 2020, construction began on the new community. Deemed "The New Isle," this 500 acre site located in Gray, Louisiana, approximately 40 miles north of the Isle de Jean Charles, features two bayous, wetlands, gathering spaces, and walking trails. Similar to the configuration on Isle de Jean Charles, each individual residential lot faces the water, with neighbors on both sides and a street behind. The new community also features shared public amenities including a market hall and community center.[29]

## Performance and Evaluation

The strategy being employed in Isle de Jean Charles was designed to be community-led and voluntary. The Biloxi-Chitimacha-Choctaw tribe has been involved in the entire planning and design process of the new community and current residents of the island are given the option to stay or go. Compared to forced relocation, through either eminent domain or disaster declarations, this approach can build trust between residents and the government and can reduce the potential for constitutional violations.

Collective moving also helps to preserve existing social networks, political systems, worldviews, and traditions. For example, with the Biloxi-Chitimacha-Choctaw tribe, there are strong cultural customs that are at risk of disappearing without a coordinated, wholesale resettlement plan. Thus, if relocation happens haphazardly, either individually or family by family, these traditions could fade.

Lastly, this approach could serve as a model for other climate-induced resettlement projects, in places outside of coastal Louisiana and even for other climate-related disasters like wildfire.

## Challenges and Future Research

This technique, though, has significant drawbacks and the fact that two resettlement initiatives have already failed for the Biloxi-Chitimacha-Choctaw tribe underlines the large challenges some communities face. Logistically, one of the biggest hurdles relates to funding. Federally- or locally-supplied funds may be insufficient or unreliable, and in some cases, may be repealed in the middle of a project. This puts a large financial burden on the community wishing to resettle. The funding issue is often tied to long timeframes needed to develop a resettlement plan. Coordinating the movement of an entire community takes years, and funding availability may shift during that time.

Another issue is a potential lack of trust of government authorities or of the resettlement process itself. If there is not constant and direct engagement with the community during all phases of the project, distrust could grow. One way to build trust is to hire local community members to work on the project and to host more workshops focused on co-designing the project.

Communities living near proposed resettlement locations may also have strong opinions about the process that could slow or even halt projects. For example, there may be a concern that resettling projects could overload existing communities with those displaced. This could lead to nearby residents or municipalities being unwilling to sell property for the new development.

Another concern is that the new location may not be able to support the original economies and livelihoods of residents. This could lead to longer commute times or even lost wages if employment opportunities are scarce or inappropriate for the community. On a related note, there is also the issue that many residents feel an intangible connection to the place where they live and it may be hard to recreate that connection in a new place.

Lastly, wholesale resettlement projects are often most successful when they are reacting to a disaster, be it floods or fires, instead of being a proactive, preventative measure. It is hard enough to convince people to leave their homes after a catastrophic natural event.

1 mi

Illilouette Creek Basin

*Fire Surrendering*

Yosemite Wilderness

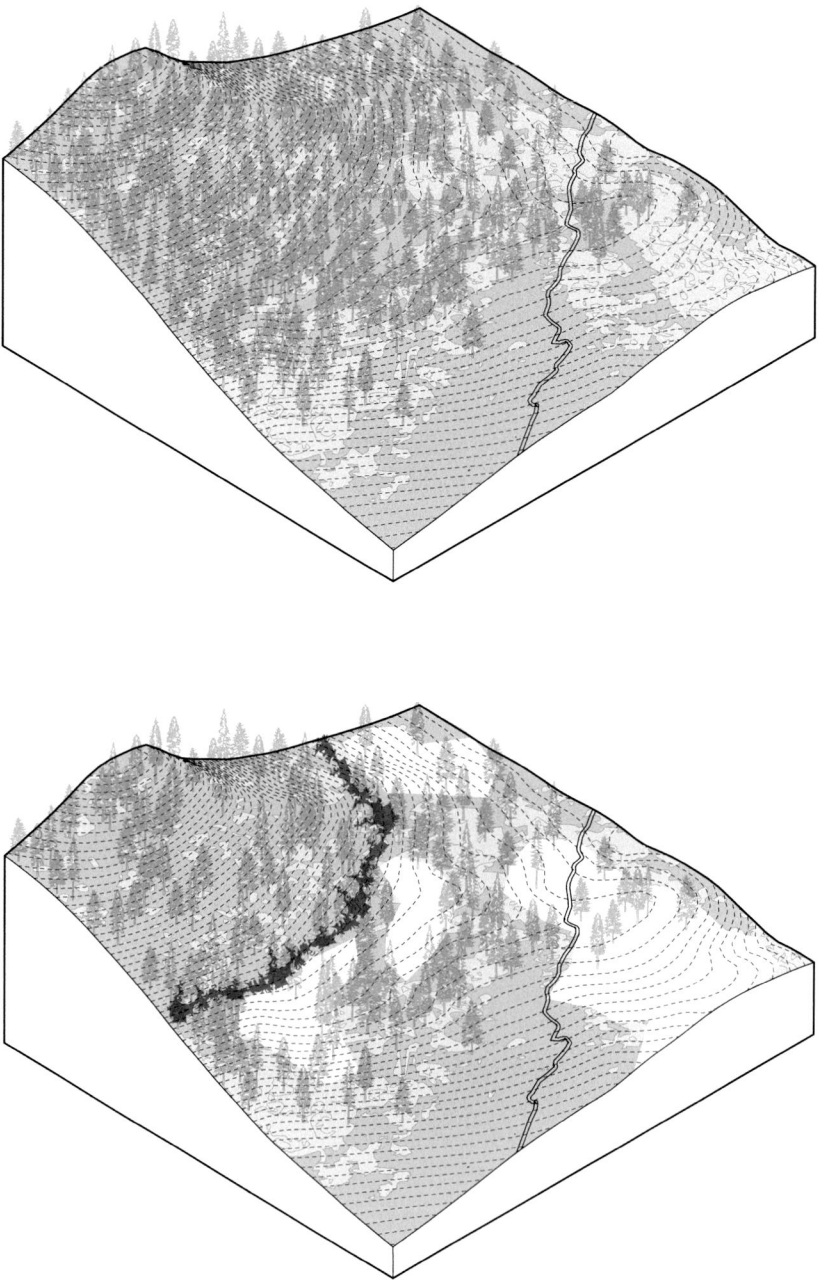

*Fire Surrendering*

| | |
|---|---|
| Description | Actively allowing the occurrence of fires on the landscape |
| Example Location | Yosemite, CA |
| Size | 40,053 acres |
| Primary Implementer | National Park Service |
| Team | US Forest Service, Stephens Lab, and the Center for Fire Research and Outreach at Berkeley Forests |

## Technique Overview

In the 1890s, fire suppression efforts began in Yosemite National Park. Then in 1911, the federal government instituted a national policy of fire exclusion, and in 1935, the U.S. Forest Service adopted a policy to put out fires the morning after detection. By the turn of the century, the number of acres burned every year in Yosemite had dropped significantly when compared to pre-suppression times.[30]

While these fire suppression efforts succeeded in protecting park infrastructure from wildfires, they also dramatically transformed its landscape. Without fire on the land, the forests of Yosemite grew into dense, homogeneous masses with high fuel loads and water needs. This shift also made the landscape more vulnerable to future catastrophic wildfire events – putting communities, structures, water supplies, and regional forest health in jeopardy.[31]

One technique being used is allowing lightning-ignited fires to burn the landscape instead of immediately putting them out upon detection. The goal of this technique is to return the landscape to a fire regime more similar to what the park was like prior to Euro-American settlement.[32]

## Planning and Design Process

This technique was initially implemented in 1972 when Yosemite National Park created a wilderness fire zone. In this zone, all lightning-ignited fires are allowed to freely burn unless they threaten nearby structures or significantly impact air quality.[33]

Figure 6.15 (*Previous*) Aerial of Yosemite. Source: Google Earth 2022.

Figure 6.16 (*Left*) A view of a fire-suppressed hillside (top) and the same view after wildfires were actively allowed to burn the landscape (bottom).

The Illilouette Creek Basin, located just southeast of Yosemite Valley in the central Sierra Nevada range of California, is part of this wilderness fire zone and has seen a dramatic uptick in mixed-severity wildfires since the 1970s. From 1930 (when Yosemite National Park began mapping fire perimeters) to 1972, only 67 acres burned in the basin with 99 lightning-ignited fires suppressed by authorities. From 1973 to 2011, nearly 20,000 acres burned in the basin from nearly 160 fires – 137 of these were lightning ignitions, 19 were accidental anthropogenic fires, and one was a prescribed burn. Over time, these wildfires burned the landscape in a patchwork-style pattern, with fires naturally extinguishing or losing speed upon entering prior burn zones.[34]

## Performance and Evaluation

The landscape of Illilouette Creek Basin is one of the only watersheds in the American West to have a less suppressed fire regime because of this technique and its successful implementation over the last several decades. Now, one part of Yosemite National Park has a landscape that somewhat resembles what the forest looked like prior to Euro-American settlement (while still notably lacking cultural burning and traditional ecological stewardship of these lands prevalent prior to that settlement). Furthermore, due to extensive research programs in the basin and decades-long monitoring programs, fire ecologists can closely study the impacts of this reintroduced fire on the landscape.[35]

Based on this work, researchers believe that letting wildfires burn in landscapes like Illilouette Creek Basin can improve forest health. Under wildfire suppression and the eradication of cultural burning, the basin had evolved into primarily a mixed conifer forest. But now it has greater habitat diversity. Forest cover has reduced by 22% while meadow areas and shrublands have increased by 200% and 24%, respectively.[36] Furthermore, the forest that remains is more heterogeneous in terms of tree size, basal area, and canopy cover.[37] Lastly, these fires have also helped support diverse, drought-tolerant wetlands that still contain moisture in the middle of summer, serving as critical habitat for a range of species.[38]

Beyond increased moisture for wetlands, the fires have also reduced plant and tree water consumption due to a decrease in competition. There has also been an increase in the basin's snowpack and a longer release of melt in the spring and summer.[39] Additionally, there is more stream flow and subsurface water storage.[40]

An increase in pyrodiversity on the landscape has also positively affected pollinator communities. It is estimated that a 5% increase in the basin has led to an increase of one pollinator and one flowering plant in the landscape. The patchwork of fires has also created more spatial heterogeneity for these communities and made them more resilient in the face of drought.[41]

Furthermore, allowing some areas of the Sierra Nevada range to burn might help to reduce fire suppression costs, which have skyrocketed in recent years. For example, between 2016 and 2020, the U.S. Forest Service spent an average of $1.9 billion every year in suppression efforts alone. This technique might also help to improve firefighter safety in the long run.[42]

## Challenges and Future Research

While returning wildfire to the Illilouette Creek Basin has succeeded on many fronts, the area is still in a fire deficit from pre-1973 suppression efforts.[43] In addition, while many fires burned in the basin over the last several decades, it is estimated that 24% of these were still suppressed due to concerns about potential property damage and reduced air quality in the region.[44] Given that this region was fire stewarded by people prior to European colonization, this *wildfire only* treatment (a retreat of human intervention) is different and likely produces different effects in the landscape, from wildfire combined with cultural burning, which is something that could be tested and explored in future research to understand differences and similarities.

There are also roadblocks that could spur suppression including being noncompli-ant with state or federal environmental acts, endangering or discouraging visitors to the park, promoting invasive alien species spread post-burn, and pulling personnel off of tradi-tional suppression duties during peak fire season.[45] So, while over 2,000,000 acres of wil-derness area in the Sierra Nevada range have been identified for this technique, more work still needs to be done to catalyze a paradigm shift in wildfire policy.[46]

One way to do this is to advocate for the restructuring of federal wildfire funds, so that more resources support risk reduction measures like letting wildfires safely burn in the landscape. Another consideration is scaling up areas for testing the technique; this is impor-tant for making an impact at the regional and state scale. The U.S. Forest Service has sug-gested working at the scale of firesheds, which are typically around 250,000 acres in size. But to do this, multi-jurisdictional coalitions and shared management goals need to be developed to make sure everyone is on the same page.[47]

Another consideration is safety – being able to ensure that a fire allowed to burn in the landscape does not get too close to inhabited areas. Illilouette Creek Basin has a natural fire-break around it – a granite perimeter – making it a perfect testing ground for fire surrendering.[48] But what about in the rest of the American West? Experts suggest focusing first on truly remote forested areas. Then, when shifting to areas closer to the WUI, spending more resources on monitoring and small-scale extinguishing efforts to reduce the risk of property loss.[49]

# Notes

1   "Steep Hillside Guidelines," City of San Diego, accessed November 1, 2022, https://www.sandiego.gov/sites/default/files/legacy/development-services/pdf/industry/landdevmanual/ldmsteephillsides.pdf
2   *Home Builder's Guide to Construction in Wildfire Zones* (Washington, DC: Federal Emergency Management Agency, 2008).
3   "Steep Hillside Guidelines."
4   "Community Wildfire Safety Through Regulation: A Best Practices Guide for Planners and Regulators," National Fire Protection Association, accessed November 1, 2022, https://www.nfpa.org/-/media/Files/Public-Education/By-topic/Wildland/WildfireBestPracticesGuide.ashx
5   Robert Olshanksky, *Planning for Hillside Development* (Chicago, IL: American Planning Association, Planning Advisory Service, 1996).
6   Ibid.
7   Ibid.
8   *Home Builder's Guide.*
9   "The Otay Ranch Resort Village Alternative H Specific Plan," accessed November 1, 2022, https://www.sandiegocounty.gov/content/dam/sdc/pds/ceqa/OtayRanchVillage13Resort/Recirc/C/1.%20%20Specific%20Plan_The%20Otay%20Ranch%20Resort%20Village_AltH_02192019%20(Optimized).pdf
10  Ibid.
11  "Petition of Sierra Club," Superior Court of California, County of San Diego, Minute Order, accessed November 1, 2022, https://oag.ca.gov/system/files/attachments/press-docs/Otay%2014%20Decision.pdf
12  Xavier Becerra, "Otay Ranch Resort Village," accessed November 1, 2022, https://oag.ca.gov/sites/all/files/agweb/pdfs/environment/comment-letter-re-otay-ranch-village-13.pdf
13  "Petition of Sierra Club."
14  Ibid.
15  "Long-Term Community Recovery Plan, Paradise, California," accessed November 1, 2022, https://www.townofparadise.com/sites/default/files/fileattachments/recovery/page/2071/6.24.19_long_term_community_recovery_plan.pdf
16  Ibid.
17  "Firebreak: Wildfire Resilience Strategies for Real Estate," Urban Land Institute, accessed November 1, 2022, https://www.preventionweb.net/publication/firebreak-wildfire-resilience-strategies-real-estate
18  "Long-Term Community."
19  "Paradise Nature-Based Fire Resilience Project Final Report," Conservation Biology Institute, The Nature Conservancy and Paradise Recreation & Parks District, accessed November 1, 2022, https://www.paradiseprpd.com/files/fcda41b0a/1.Paradise.Final.Report.2020.0715.pdf
20  Ibid.
21  Ibid.
22  Ibid.
23  Ibid.
24  Ibid.
25  Madaline King, "A Tribe Faces Rising Tides: The Resettlement of Isle de Jean Charles." *LSU Journal of Energy Law and Resources* 6, no.1 (2017): 295–317.
26  Ibid.
27  Jessica Simms, Helen Waller, Chris Brunet and Pamela Jenkins, "The Long Goodbye on a Disappearing, Ancestral Island: A Just Retreat from Isle de Jean Charles." *Journal of Environmental Studies and Sciences* 11 (2021): 316–328.
28  King, "A Tribe Faces."
29  "Community Master Planning and Program Development for the Isle de Jean Charles Resettlement: Phase 2 Report," accessed November 1, 2022, https://isledejeancharles.la.gov/sites/default/files/public/IDJC_Phase2Report_7–21-21.pdf
30  "Confronting the Wildfire Crisis: A Strategy for Protecting Communities and Improving Resilience in America's Forests," USDA Forest Service, accessed November 1, 2022, https://www.fs.usda.gov/sites/default/files/Confronting-Wildfire-Crisis.pdf
31  Rachelle Hedges and Gabrielle Boisramé, "Illilouette Creek Basin Research and Publications," Stephens Lab, accessed November 1, 2022, https://forests.berkeley.edu/sites/forests.berkeley.edu/files/Illouette_Creek_Basin_Summary_2020_Update_Final.pdf
32  Scott Stephens et al., "Final Report – Hydrology and Fire in the Sierra Nevada: A possible Win-Win," accessed November 1, 2022, https://www.firescience.gov/projects/14-1-06–22/project/14-1-06-22_final_report.pdf
33  Jan van Wagtendonk, Kent van Wagtendonk, and Andrea Thode, "Factors Associated with the Severity of Intersecting Fires in Yosemite National Park, California, USA." *Fire Ecology* 8, no. 1 (2012): 11-31.
34  Ibid.
35  Hedges and Boisramé, "Illilouette Creek."
36  Stephens et al., "Final Report."
37  Water Resources.
38  Gabrielle Boisramé, Sally Thompson, Christina Tague, and Scott Stephens, "Restoring a Natural Fire Regime Alters the Water Balance of a Sierra Nevada Catchment." *Water Resources Research* 55, (2019): 5751–5769.

39   Ibid.
40   Hedges and Boisramé, "Illilouette Creek."
41   Ibid.
42   "Confronting the Wildfire Crisis."
43   Wagtendonk, Wagtendonk, and Thode, "Factors Associated."
44   Stephen Pyne, "Pyrocene Park," *AEON*, March 24, 2022, https://aeon.co/essays/what-yosemites-fire-history-says-about-life-in-the-pyrocene
45   Wagtendonk, Wagtendonk, and Thode, "Factors Associated."
46   Stephens et al., "Final Report."
47   "Confronting the Wildfire."
48   Pyne, "Pyrocene Park."
49   "Confronting the Wildfire."

# Bibliography

Becerra, Xavier, "Otay Ranch Resort Village." Accessed November 1, 2022. https://oag.ca.gov/sites/all/files/agweb/pdfs/environment/comment-letter-re-otay-ranch-village-13.pdf

Boisramé, Gabrielle, Sally Thompson, Christina Tague, and Scott Stephens, "Restoring a Natural Fire Regime Alters the Water Balance of a Sierra Nevada Catchment." *Water Resources Research* 55, (2019): 5751–5769.

"Community Master Planning and Program Development for the Isle de Jean Charles Resettlement: Phase 2 Report." Accessed November 1, 2022. https://isledejeancharles.la.gov/sites/default/files/public/IDJC_Phase2Report_7-21-21.pdf

"Community Wildfire Safety Through Regulation: A Best Practices Guide for Planners and Regulators," National Fire Protection Association. Accessed November 1, 2022. https://www.nfpa.org/-/media/Files/Public-Education/By-topic/Wildland/WildfireBestPracticesGuide.ashx

"Confronting the Wildfire Crisis: A Strategy for Protecting Communities and Improving Resilience in America's Forests," USDA Forest Service. Accessed November 1, 2022. https://www.fs.usda.gov/sites/default/files/Confronting-Wildfire-Crisis.pdf

"Firebreak: Wildfire Resilience Strategies for Real Estate," Urban Land Institute. Accessed November 1, 2022. https://www.preventionweb.net/publication/firebreak-wildfire-resilience-strategies-real-estate

Hedges, Rachelle and Gabrielle Boisramé, "Illilouette Creek Basin Research and Publications," Stephens Lab. Accessed November 1, 2022. https://forests.berkeley.edu/sites/forests.berkeley.edu/files/Illouette_Creek_Basin_Summary_2020_Update_Final.pdf

*Home Builder's Guide to Construction in Wildfire Zones.* Federal Emergency Management Agency, 2008.

King, Madaline, "A Tribe Faces Rising Tides: The Resettlement of Isle de Jean Charles." *LSU Journal of Energy Law and Resources* 6, no.1 (2017): 295–317.

"Long-Term Community Recovery Plan, Paradise, California." Accessed November 1, 2022. https://www.town-ofparadise.com/sites/default/files/fileattachments/recovery/page/2071/6.24.19_long_term_community_recovery_plan.pdf

Olshanksky, Robert. *Planning for Hillside Development.* American Planning Association, Planning Advisory Service, 1996.

"Paradise Nature-Based Fire Resilience Project Final Report," Conservation Biology Institute, The Nature Conservancy and Paradise Recreation & Parks District. Accessed November 1, 2022. https://www.paradise-prpd.com/files/fcda41b0a/1.Paradise.Final.Report.2020.0715.pdf

"Petition of Sierra Club", Superior Court of California, County of San Diego, Minute Order. Accessed November 1, 2022. https://oag.ca.gov/system/files/attachments/press-docs/Otay%2014%20Decision.pdf

Pyne, Stephen, "Pyrocene Park," *AEON*, March 24, 2022. https://aeon.co/essays/what-yosemites-fire-history-says-about-life-in-the-pyrocene

Simms, Jessica, Helen Waller, Chris Brunet and Pamela Jenkins, "The Long Goodbye on a Disappearing, Ancestral Island: a Just Retreat from Isle de Jean Charles." *Journal of Environmental Studies and Sciences* 11 (2021): 316-328.

"Steep Hillside Guidelines," City of San Diego. Accessed November 1, 2022. https://www.sandiego.gov/sites/default/files/legacy/development-services/pdf/industry/landdevmanual/ldmsteephillsides.pdf

Stephens, Scott, John Battles, Maggi Kelly, Sally Thompson and Brandon Collins, "Final Report – Hydrology and Fire in the Sierra Nevada: A Possible Win-Win." Accessed November 1, 2022. https://www.firescience.gov/projects/14-1-06-22/project/14-1-06-22_final_report.pdf

"The Otay Ranch Resort Village Alternative H Specific Plan." Accessed November 1, 2022. https://www.sand-iegocounty.gov/content/dam/sdc/pds/ceqa/OtayRanchVillage13Resort/Recirc/C/1.%20%20Specific%20Plan_The%20Otay%20Ranch%20Resort%20Village_AltH_02192019%20(Optimized).pdf

van Wagtendonk, Jan, Kent van Wagtendonk, and Andrea Thode, "Factors Associated with the Severity of Intersecting Fires in Yosemite National Park, California, USA." *Fire Ecology* 8, no.1 (2012): 11–31.

# Epilogue

# Pyro Futures

Figure 7.1

View of a landscape around Lake Berryessa after the LNU Lightning Complex Fires of 2020.

# Chapter 7

# Pyro Futures

## Introduction

One certainty we have, is that future landscapes of the world's fire-prone regions will be different from what they are today, and we know that we don't know what they will become with any certainty. We live in times of accelerating change across technological, ecological, climatic, economic, cultural, and socio-political domains, and fire is a key factor of landscape change across them all.

In the previous chapters, we examined three different approaches to fire: resistance, co-creation, and retreat. In doing the research for this book, it became clear that new and revived approaches for working intentionally, creatively, and proactively with fire are expanding, and we feel that co-creative approaches are where our human-fire coevolution has historically been, and where we are likely to return.

Why? Resistance techniques of fire suppression have proven to be a failure and a mistake; a massive colonial mismeasure, which, for future generations, may appear as a 150-year anomaly in the larger context of human-landscape evolution with fire. Experts and managers have known for decades that fire suppression doesn't work, and it's taken about 50 years for that course correction to culturally establish itself. Fire proofing structures and aggressively extinguishing wildfires near developed land will continue to have its place, but striving to eradicate fire from landscapes where it is an integral and regenerative force is something we now know doesn't work out as intended. Rather, resistance and fire suppression create a *fire-fighting trap* that serves to create more severe and damaging fires.[1] We need a new paradigm of proactive, co-creative, humble, and forward-thinking approaches. Comparatively, retreat will be necessary from the increasingly vulnerable places where people are building and rebuilding communities, such as the vast, transitory domains of the wildland-urban interface (WUI). But wholesale retreat, or fully withdrawing from landscapes and leaving them untended and unmanaged, may be just as detrimental as resistance. Many of these landscapes have been deeply altered by logging, the introduction of non-native species, and growing impacts of climate change. Without adaptive management, rural and feral landscapes might develop in ways that are undesirable and detrimental. Retreat also runs contrary to all we know of how extensively landscapes such as California's were tended to by native tribes – and how biodiverse and aesthetically cherished those lands were through that stewardship – prior to colonization.[2] California's most healthy landscapes have long been ones co-evolved with people.

DOI: 10.4324/9781003172956-10

In truth, all three approaches to fire curated in the previous chapters are co-creative and coevolutionary, but in radically different ways that engender very different worlds. A resistant effort to fully control fire, has largely served to make fire act more fiercely and uncontrollably. Retreat does the exact opposite, leaving fire to run wild as it will, but with little acknowledgement of the highly altered, and often unhealthy state in which we would be leaving these lands, and with little foresight or planning for what they will become. This kind of walking away – the withdrawal from the post-colonial landscape – inscribes very particular evolutionary pathways, none of them being "natural."

As some have observed, "what becomes of wildfire next is anyone's guess."[3] Wildfire, as a "strikingly immediate object, morphs into an array of disruptive unknowns."[4] With so many indeterminate factors affecting what future fires will be, there is no clear, definitive future we can rely on. Our future of accelerated change can be so many different things, and much of what becomes will depend on the decisions and actions societies choose to make now, that address the novel legacies we have inherited.

Runaway change, according to anthropologist Adriana Petryna, is "the rapid departure from long-established baselines, systemic patterns, or historical trends," which makes reliable future prediction, control, and projection untenable.[5] Runaway change enhances future indeterminacy, and in the face of accelerating ecological crises, climate destabilization, and socio-political inequities, it's tempting to throw our hands up in resigned apathy. But that can easily make things worse, and squander the actionable time we have to shape and guide our collective future. As Petryna states:

> One can also think of runaway change as signifying a gulf between what is expected and what actually occurs: an empty dimension that might not be so empty after all, but rather an elaborately staged absence, more like a vacuum, in which certain knowledge of the thing itself—wildfire, for example—continually disappears. If this is the case, namely, that runaway change conceals within itself the impetus for its own progression, then what does it take to render this change observable and, if possible, to slow it down? How are temporal horizons being rendered with respect to runaway change? How can actionable time be recovered or pried away from the pace of its advance? What would it mean to gain ground in this peculiar race against (lost) time, and what is at stake for agency and retreat in the negotiation of surprise?[6]

Without doing the hard work of imagining the many different kinds of futures that may come to be, we risk being blindsided and finding ourselves enmeshed in oncoming realities we likely don't want. To affect the future, we, as a diverse and often conflictual society, need to do the challenging work of imagining what worlds might actually manifest and how we might want to affect those manifestations. This is the intellectual and ethical labor of horizoning work: of "recognizing and pushing against the limits of what is known" to imagine what might lie beyond our immediate reality.[7] We need imaginations and forms of design research that brings runaway futures into the present as an object of inquiry, experimentation, and active intervention.

In this chapter, we return to California, and rather than exploring its past, we investigate its pyro futures and how the pathways to those futures might be transitioned in more just and intentional ways.

## Paths of Change

Horizoning work requires thinking like a futurist and imagining different realities that could come to be. It requires creative speculation on how the world might evolve and change. Scenario planning is one disciplined way of going about this, wherein one seeks to identify key factors, trends, and drivers of change in a current situation, and then explore how these trends and drivers might shift over time, both on their own terms and in relation to one another.[8] The goal of such design speculation is not to exhaustively imagine every possible future. Rather, if we can imagine a small and diverse plurality of futures, it can enhance our ability to be more selective about the future we want, and to more strategically plan and act to try to make it real. As a specific place, California has unique drivers of change, but these may also serve as factors in other fire-prone regions. We begin with describing what we see as the most critical and interrelated factors to consider in the future of fire.

### *Global Warming and Accelerated Landscape Change*

The effects of human-induced climate change are being experienced now, and these effects will certainly increase in magnitude over time. How fast and severely climate change happens, and how it will alter landscapes in complex and specific ways, is uncertain. Right now, we can't know if global carbon emissions will ever be significantly reduced through governance or other means, and even if we did know what future atmospheric carbon levels will be, only guesses can be made as to the way the earth will respond.

But for fire specifically, overall trends due to climate change are clear. In a warmer climate, fire begets more fire. Fire regimes are poised to accelerate, trending toward shorter intervals between pyric events, combined with longer fire seasons in which this can happen. Increased drought severity, drier soils, and more climate volatility will place greater stress on forests and many plant communities, thus increasing mortality and creating more fuel for wildfires.[9] Higher velocity wind patterns and a more volatile atmosphere in general will allow for wildfires to spread faster and further, thus making fire behavior less predictable and less manageable.

Fire, and the ecological conditions it changes and creates, has been deeply affected by human actions since we, as upright, bipedal mammals with large brains, have been present in sizable numbers. Anthropogenic climate destabilization (mainly through the mining and combustion of fossil fuels) takes this to a novel, globalized scale. And when combined with other human influences, such as increased ignitions, introduced plant species, logging, and other types of extractive land management, ecological associations and habitats are rapidly transitioning. It has been estimated that as many as 95% of California's wildfires are started by human activity.[10] In California's wildlands we are already witnessing wholesale shifts in ecological plant communities, like oak woodlands and conifer forests turning into non-native grasslands, due to too frequent fire events.

Accelerated climate change accelerates landscape change in complex ways that often creates aggregating and cascading effects. For example, in California we are expected to see decreasing snowpack in the Sierra Nevada mountains, which has already been reduced by 23% since 1955,[11] and is projected to be reduced by 79% by 2100.[12] With these changes, winter precipitation will literally flow through California's terrain much faster, with much less being absorbed, which in turn, will lead to drier soils and longer, intensified summer drought conditions. This is leading to increased tree and other plant mortality in places like the Coastal Ranges and Sierra Nevada Mountains, which creates more <u>fuels</u> for fire. If we combine this change with groundcover dominated by non-native grasses (which is considerably more flammable and has more biomass than native meadows) combined with fire suppression techniques, we get the exemplary conditions for mega-fires.[13] Many of these fires are of such high intensities that native, fire-adapted trees, like giant sequoias, can't survive them.[14] Combustion temperatures are so high that these fires burn just about everything, reducing soil carbon and killing off its microbes and biota. And given that the scale and extent of high intensity fires has been expanding, the patch size of these effects is huge and unprecedented. Landscapes subjected to these fires are radically transformed, and when precipitation does return to these lands (as increasingly more intense rain storms) it causes erosion and stream sedimentation rates that exceed historical precedents by orders of magnitude. The degree to which landscapes can recover from these impacts is unknown, but many are likely to convert to some new assemblage or biome that is different from what they were prior to these kinds of fires.[15]

California encompasses a remarkable range of habitats and biodiversity, each of which are likely to change in different ways. What we know is that all of them are subject to novel conditions and are changing rapidly. Formative landscape rhythms and cadences – such as <u>fire regimes</u> and weather patterns – have mutated, creating a range of new temporalities and asynchronies. All of these climatic and space-time changes are closely interrelated and affect all the other fire factors listed below.

## *Property and Land Use*

Land use is the strongest determinant of the forms, development, activities, and types of management that occur upon landscapes. Ownership, particularly public or private, is a major factor in what kinds of uses and management protocols are put in place. Public lands, such as parks and wilderness areas, are often managed quite differently and often at larger scales than private lands. Across both public and private land, there is a growing priority to manage wildfires more proactively, more widely, and more sustainably.

As we've discussed, the <u>wildland-urban interface</u> is the fastest growing land use in the state of California, and is also the most vulnerable to wildfires. Approximately one third of all Californians currently live in the disparate and widely dispersed WUI.[16] The WUI is not a fixed geographic location, rather it is a shifting and changeable condition in which the edge of urbanization meets the wild in <u>combustible</u> ways.[17] Faced with increasing wildfire threats, will the WUI continue to rapidly expand as it has over the past few decades, or will that trend slow, or even stop? Will we continue to rebuild communities that catastrophically burn, or might we strategically retreat? From our research looking across multiple trends of change, we think that slowing of WUI expansion is inevitable, and retreat ever more likely as we progress into a hotter climate. When and how that withdrawal happens raises many socio-cultural and socio-political questions. Whether that urbanized retreat will be planned, safe and equitable, or delayed, forced and catastrophic remains to be seen.

The very notion of property, and the individualistic and colonial assumptions built into it, may be loosened in the future, as greater cooperation, communality, and co-management of landscapes are required to effectively co-exist with fire, particularly in the WUI.

Figure 7.2 (*Right*) A map of California showing the wildland-urban interface extents (in grey) and the historical fire perimeters (in orange).

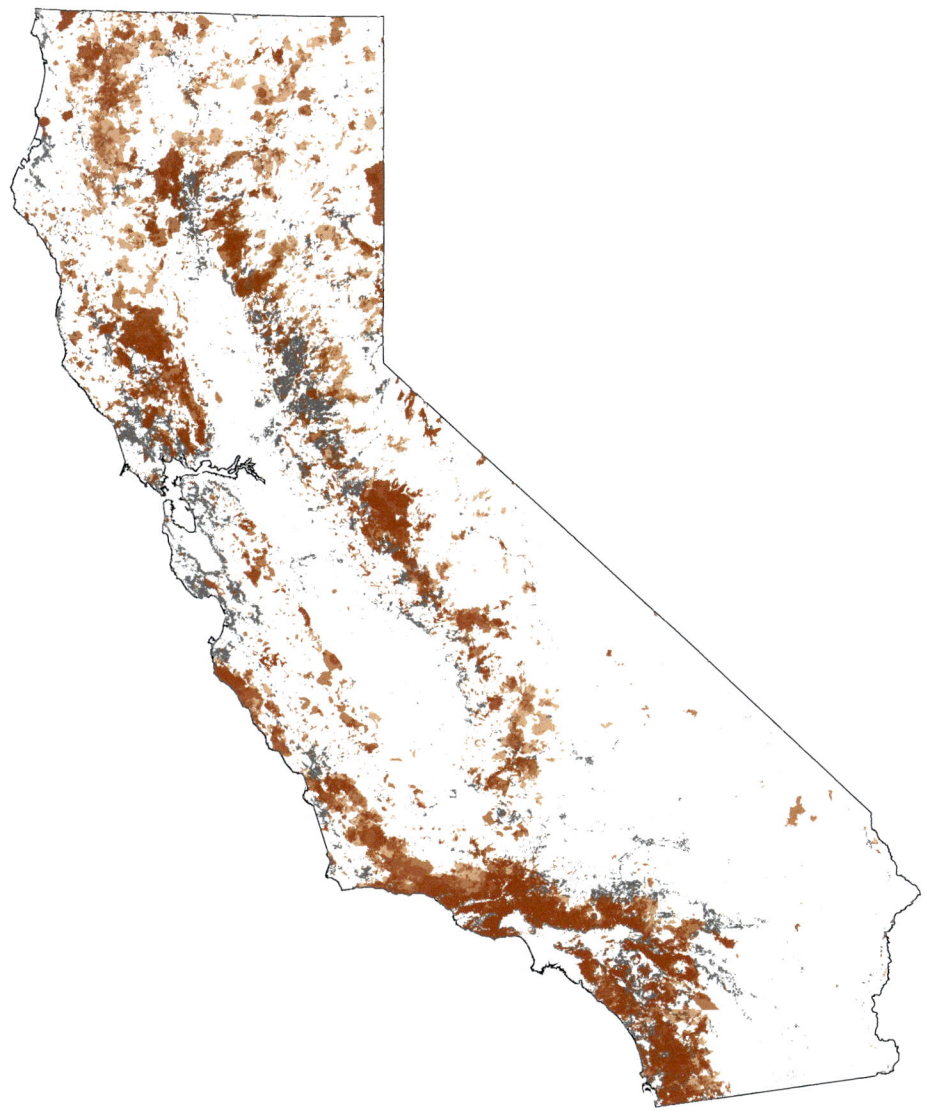

## Management and Stewardship

Researchers have estimated that approximately 4.5 million acres of California burned annually prior to the 1800s.[18] California was historically fiery and smoky. Since colonization, far less area of the state burns per annum. Even though there are increasing numbers of mega fires (fires that burn more than 100,000 acres of land), the overall area burned is far smaller than the pre-colonial average. As we have documented in the *co-creation* design techniques, many people are trying to get more fire on the landscape more frequently and beneficially through active stewardship, including prescribed burns, cultural burns, and strategically managing wildfires. As we've (re)learned, the way to nurture a forest and other fire-dependent landscapes is by burning them.[19] This is the nurturing of "good" fire (rather than "bad"), and the cleansing, regenerative qualities it can generate. These methods can be combined and supplemented with fire surrogacy techniques, such as mechanical forest thinning, selective land clearing, conservation grazing, and changing land uses and productive land covers.[20]

In California and elsewhere, there is no real debate of the need for increasing the presence and application of beneficial fire.[21] The key questions center around how it's performed, how to do it at the scale that it is needed, and how to do it in a manner that can be sustained over time. *California's Strategic Plan for Expanding Beneficial Fire* sets a target of scaling up the application of beneficial fire to 400,000 acres annually by 2025.[22] This would be a major increase from current levels of intentional burning, but is a long way away from the 10–30 million acres of land in the state in need of proactive fire management.

The main constraints on expanding fire management in California (both fire and fire surrogates) are the labor, time, and money that such work requires, and for intentional burning specifically, numerous risk factors, liability concerns, and cumbersome policy and permitting requirements that must be navigated.[23] State government has begun to dedicate more funds to fire stewardship, but will funding be adequate to meet the scale of the effort required, and to maintain it? Who might this new workforce consist of? Will they be diverse and equitably compensated? How and by whom will they be trained? Given the pressing need to create a new skilled employment sector that can potentially save lives, improve public health, improve ecosystem health and services, and save money,[24] there is much opportunity for creative investment and strategic planning in this domain, ranging across state and federal initiatives and funding mechanisms.

There is also the question of which kinds of land-fire stewardship techniques will be advanced and practiced, for as we have shown, there are diverse ways of designing with and for fire, each with their own affordances and potentials. Continued on-the-ground experimentation and prototyping is the most sure way to refine these approaches and foster greater application.

## Ecocultural Restoration and Indigenous Sovereignty

Legacies of indigenous land dispossession have played a defining and definitive role in why wildfires have become so problematic and destructive. In colonizing California, and in deliberately suppressing traditional ecological knowledge and forms of land stewardship, the state's landscapes have been eco-culturally transformed. This change is most pronounced with respect to fire setting. Indigenous cultures were, and are fire-dependent,[25] in contrast to a colonizing presence that was fire phobic. Fire was the First Nations' most essential and impactful tool in designing and stewarding multifunctional and multivalued landscapes.[26]

In changing contemporary Western relationships with fire, there is also a need to address colonial missteps. Returning fire to California's landscapes should not just be eco-logically restorative, but eco-culturally restorative. Cultural burning, as practiced by indige-nous tribes, is typically quite different and more multidimensional in design than prescribed burns. In design intent, they go far beyond fuel reduction, often timed to produce targeted, desired effects on specific species and habitats and are typically smaller in scale and of lesser fire intensity than the heavier handed, mechanized techniques of prescribed burns.[27] Cultural burns also have sacred and animistic qualities, and are often intergenerational events (elders and children are often present) with more festive and ceremonial dimen-sions. As Robin Kimmerer and other indigenous scholars have discussed, most indigenous cosmologies are structured on reciprocity between people and land; of people being intrin-sically part of the land, thus stressing mutual benefit or symbiosis, rather than extraction, which is a profoundly different world view than contemporary American neoliberal capitalism.

Some of these differences in ways of burning may blur in the future, as cultural burning practices by tribes and cultural fire practitioners are expanding in California and gaining public recognition and support (such as the prescribed fire training exchanges we saw in chapter five's *fire lighting*). New partnerships are being forged between tribes and cultural fire practitioners with local, state, and federal government and NGOs. As the value of traditional ecological knowledge becomes increasingly recognized, that knowledge is being shared with others working to improve fire-landscape management.

How these collaborative relationships evolve and the equity they may exhibit remains to be seen, as does the degree to which they address colonial legacies or perpetu-ate them. The expansion of cultural burning does continue to face challenges, such as lack of understanding or respect for traditional fire stewardship by non-indigenous people, cum-bersome permitting processes, and lack of access to ancestral homelands.[28] There are also more fundamental questions of indigenous sovereignty in general, such as indigenous rights to burn on their own terms, rather than those of state governance.[29] Why should tribes have to get permission to burn on lands that were taken from them? In the words of cultural burning practitioner Margo Robbins (interviewed in chapter five), indigenous sover-eignty needs to be more fully recognized so they "can burn in the right place, at the right time, for the right reasons."[30] And as the writers of The *Red Deal* explain, the only real way to undo the injustices of land colonization is to give land back to those from whom it was taken.[31] The degree to which decolonizing can or will happen with cultural burning – through indigenous leadership, co-management, or full land management rights, and giving land back – is uncertain, and closely tied to law and policy.

## Insurance and Liability

When conducting a prescribed or cultural burn, or letting a wildfire created by lighting run its course, there is always the possibility that the fire might burn differently or over more area than planned, or get out of control. If it does, it may damage private property or cause unhealthy air quality conditions, due to the smoke. This creates liability issues for those trying to beneficially work with, and sustainably manage fire.

Prescribed fire liability insurance protects the people and organizations conducting prescribed fires, should claims be made against them related to fire or smoke, or due to more nebulous claims, like changes in landscape views, property values, or impacts to wildlife. With the increasing risks of wildfires caused by past mismanagement and climate change impacts, insurance has (through reinsurance protocols) increased greatly in cost and has become more challenging for fire stewards to acquire. This creates impediments to intentional burning that are most problematic for private entities, landowners, and cultural burners. As Daniel Godwin of the *Ember Alliance* writes,

> *Private entities create a public good by reducing wildfire risk and protecting communities, but many are forced to internalize all the risk. Without prescribed fire liability insurance, these organizations expose themselves to existential risk to their finances and their operations and, understandably, many choose not to engage at all.*[32]

Many beneficial burners are getting caught in a catch-22, where the need for intentional burning is increasing and gaining acceptance, but the personal risk fire workers are forced to assume is also increasing, thus limiting burning, which in turn, continues to increase public fire risk.

Property insurance is another key factor, particularly in high fire risk places like the WUI. Generally, insurance costs are rising across many sectors due to broadly increased environmental risks (be it sea level rise, river flooding, wildfire, drought, and other aggregating risk factors). Understandably, it has become more expensive to insure properties in high- risk areas due to more claims. Increasingly, insurance might not be offered for some regions or states at all, which might be more equitable or appropriate than it seems. There are ethical questions here concerning if property insurance should be offered at all in high- or very high-risk areas, given that costs to rebuild crossover into the public domain, and the promise of being able to safely or responsibly rebuild is becoming an increasingly false one, with impacts hitting far beyond claimants. Reductions to, or the elimination of property insurance may end up being an instrument of retreat from dwelling in increasingly risky areas, whether that transition is inadvertent or strategic.

How insurance and liability concerns and protocols evolve with respect to wildfire is unknown and will significantly influence future land development and fire stewardship. How U.S. culture and governance protocols understand and approach risk is in transition generally, which is more obvious in flood risk and national revisions to flood policies and infrastructures. Much of this transition is based on the understanding that we cannot "control" risk, and are living in an increasingly risk-prone world of our co-making.

## *Investments, Policy, and Governance*

Past policies in California and the U.S. were largely structured to resist and suppress wildfires, suppress cultural burning, and to allow for or incentivize development of the WUI, which have collectively brought us to the challenging time and place we are now. Over the past 50 years, the dysfunction of these approaches has become increasingly acknowledged in tandem with the expansion of the problems encountered. And there is nothing more effective than disasters and breakdowns to spawn evolution in policies and protocols.

In response to the wildfire crisis, California policy and governance and Federal policies are showing signs of proactive change to try to address it. Funding for wildfire management and research has increased in recent state budgets, and there are new strategic plans and legislation targeting wildfire management. For example, Senate Bill 332 (passed in 2021) states that The CA Legislature "finds and declares that in order to meet fuel management goals, the state must rely on private entities to engage in prescribed burning for public benefit."[33] This Bill provides assurances that:

> no person shall be liable for a prescribed burn if specified conditions are met, including, that the burn be for the purpose of wildland fire hazard reduction, ecological maintenance and restoration, cultural burning, silviculture, or agriculture, and that, when required, a certified burn boss review and approve a written prescription for the burn.[34]

This bill limits liability[35] and explicitly codifies what *cultural burns* and *cultural fire practitioners* are (as distinct from prescribed burners) and exempts them from some prescribed burn requirements, including burn bosses, burn plans, and the need for written permission from the state to burn.

Assembly Bill 642 (also passed in 2021) requires the Director of California's Department of Forestry and Fire Protection (CAL FIRE) to identify areas in the state as moderate and high fire hazard severity zones and make this information publicly available (previous law only required the CAL FIRE to identify areas that were considered very high fire hazard severity zones).[36] These far more expansive fire severity zone maps are now openly available online. AB 642 also requires the CAL FIRE's director to appoint a cultural burning liaison to the State Board of Fire Services to advise their agency on developing increased cultural burning activity, engage in recruitment efforts with California Native American tribes and tribal organizations, and for cultural fire practitioners to fill vacancies within its department. Among other initiatives, the Bill requires CAL FIRE to focus on public education efforts for fire prevention and public safety and efforts at restoring beneficial fire and cultural burning with California State Universities, California Native American tribes, tribal organizations, and cultural fire practitioners.

These and other new legislation passed in California reveal major realignments in priorities that push hard for more beneficial burning and for elevating the role of traditional ecological knowledge and the rights of California's Indigenous tribes. What is yet to be seen is, if and how these policies are fully carried forward and the ways in which they are put into practice.

As mentioned earlier, we find that retreat from increasingly high-risk areas is inevitable and necessary. Yet, most policies seem to still assume that the dispersed and extensive WUI can be sustained or safely managed, which seems to be a form of denial of impending realities. Generally, the U.S. has a questionable track record in developing just transitions for people living in harm's way, many of whom were incentivized or economically forced to do so in the first place. Creative programs and funds are needed to relocate communities in ways that are equitable and allow for self-determination. We need to develop and implement such programs to attenuate what might otherwise be a catastrophic retreat.

## Sensing, Access, and Education

This last factor is primarily concerned with how fire and fire stewardship are experienced, known and perceived by people and publics, including experts, scientists, land managers, stakeholders, institutions, and upcoming generations. Fire-dependent landscapes can be sensed in very different ways, ranging from immersive bodily encounters, such as tending a prescribed burn, or being confined indoors for weeks at a time due to toxic, smoky air, to more conceptual and abstract experiences, such as sensationalized, televised news reports of fire events. How we experience a landscape conditions and informs our perception of it, which in turn, can lead to different responses and changes in relationships.[37] For example, technologies for sensing landscapes in ways beyond what our bodies can on their own – such as heat, moisture, and air quality sensors, or drones equipped with LIDAR and multi-spectral imaging capacities – have steadily advanced and proliferated over the past decades. These technologies offer an expanded range of aesthetic access to landscape phenomena. How we will use and learn from this sensing (and who has access to that sensing) is the subject of much fire-related research.

How the general public perceives fire and fire stewardship depends on many things, such as whether or not they are actively involved in land-fire stewardship. Many people will not have that experience, such as those living in city centers. Their access to fire will likely depend on what they read, and what their children are taught in school. Education and outreach can play a major role in improving our relations with fire.

Perceptions of fire are changeable. If nearly an entire nation came to believe that fire was bad and preventable because a deep-voiced cartoon bear told them so, surely there is much opportunity to better inform and engage publics in intentionally changing our relations with fire.

## Worlds that Might Be

All the factors discussed in the previous section are inseparable from one another. They act on their own terms and in relation to what is happening in all the others. Collectively, there are nearly infinite variations in how these factors could play out and generate very different future realities. But, if we combine them in just a handful of differing ways, tweaking the knobs of each factor, we might explore a few informative future scenarios. Looking at a few horizons can help us feel into a multitude of others.

To start, we might try imagining a future where things mostly continue along current trends, such as this:

## Wrath of Fire

*Over the next two decades, little is done to curb carbon emissions at the global level. Governments variously try to do so, but efforts are largely unsuccessful in the context of declining economic gains and squabbles over which nations should do what. In tandem, climate change effects begin to hit hard, and unfortunately, the scientific projections of those impacts prove to be woefully underestimated, as they often have been. In California, which already had a remarkably variable and erratic climate prior to global warming, weather events become commonly extreme. Elevated temperatures and longer summer droughts further place many ecosystems on the brink, with alpine forests taking the worst hit. Trees in both the coastal and Sierra Nevada ranges expire in multitudes. Further expansion and development of the WUI continues to occur across a variety of forest and chaparral landscapes, mostly unchecked by laissez-faire pro-development land use policies. Proactive fire management by public agencies increases across the state, and some token moves are performed to expand indigenous* <u>cultural burning</u>*, but only modestly in comparison to the need.*

*In the summer of 2040, a forceful weather event with multiple thunderstorms sets fires burning across large swaths of the state, encompassing all mountain ranges and their lower flanks. The scale of the inferno is far beyond what CAL FIRE and other fire-fighting agencies can even imagine controlling. Massive numbers of rural communities, both those of the wealthy and the vulnerable, are destroyed. The intensity of the fires kills just about everything they burn through. Toxic smoke – filled with the tiny particulates of former houses, wires, infrastructures, and automobiles – blankets much of the state's skies for nearly two and half months. And just as it seems the fires are fading, a second storm with more extreme winds ushers in the second wave of the inferno. These fires don't fade until early December, when light rains make active containment possible.*

*Such a vast extent of the state's WUI is burned in the fires – including thousands upon thousands of structures – that it's impossible to rebuild it. Private insurance companies fold under the weight of the claims and ask to be bailed out by the federal government, which refuses, given so many other environmental crises it's dealing with. The state also discontinues its property insurance stopgaps, by necessity. And after the trauma that so many residents have experienced, many don't want to rebuild or go back anyway. Former tourist destinations, like Lake Tahoe, now have little to offer. The public health toll from the air pollution from these fires (and others before them) is experienced over time through higher rates of asthma, respiratory illnesses, and cancers.*

*People come to speak mythically about the summer of 2040 – calling it "the great retreat," or "the wrath of fire" (rather than The Grapes of Wrath) – when the state radically changed course, went bankrupt and its population plummeted and emigrated elsewhere. Huge expanses of deeply burned land are left to develop in novel, post-apocalyptic ways.*

As this scenario illustrates, the future of California could be quite bleak if we continue along the status quo. But what might happen if more proactive and adaptive measures are taken now and into the future? What if we worked harder to beneficially live with fire, and if we also got a little lucky?

## Pyric Commons

*In the late 2020s, the climate movement became an exceedingly powerful political force. Based on numerous and patchy international crises – floods, droughts, deadly heat waves, ecological and agricultural breakdowns, and famine – a motley global public often led by disillusioned and activated youth, is, through shutdowns, walkouts, sit-ins, economic embargoes, and isolated acts of eco-terrorism, able to pressure and demand that* governments immediately curb carbon emissions, or else...[38] Even the U.S., which just happens to have a progressive president at that particular four-year moment, agrees to this, which tips the scales. The Green New Deal becomes real.

This change leads to massive public and private investment in more sustainable technologies, policies, and lifestyles in the U.S. California is already a leader in this domain, and gets to work. Particular to fire management (and its proactive efforts on climate change) the state's government and policy makers make sweeping changes. Realizing the public cost of development in the WUI, and that public agencies can no longer afford or feasibly manage these areas, the urban edge becomes creatively regulated and reimagined. For communities living in very high- and high-risk zones, the state pioneers the offering of collective buyouts, wherein those communities are given opportunities to move to safer locales together, rather than on their own. This is paired with a federally supported retreat strategy for vulnerable populations who cannot afford to move out of the WUI. These managed retreats are a test to see if, over time, more money and lives are actually saved by relocating, rather than rebuilding in harm's way.

For rural communities with lower risk levels who wish to remain in place, they must create their own "community fire and landscape stewardship plan," aimed at fostering greater autonomy and self-determination. This takes "fire adapted communities"[39] to an entirely different place that is far less vague and decentralizes fire management across them. As part of these plans, residents and private landowners are provided training in pre-scribed burning (performed within the safer, off-season winter months) and these communities must actively manage and burn their lands and large extents of surrounding lands at agreed upon recurring time intervals. Legislation from the early 2020s is successful in building partnerships between tribes and State and Federal agencies, which refines the techniques, timing, and site-specific application of these burns. Community members must either directly contribute their labor to these stewardship efforts or pay their contribution. These efforts are combined with fire surrogate strategies, such as mechanical thinning (integrated with local wood product and biofuel economies), and productive land use/land cover conversion in peripheral landscapes adjacent to development. These WUI communities start to be colloquially called CSFS – community-supported fire stewardships. Over time, this creates an entirely new sector of public-private land managers that greatly diversifies this work force, restores ecological health and function to WUI and beyond WUI lands, and greatly reduces reliance on public agencies to manage and protect these communities.

Given the state's proclivities to burn, some rural communities are still occasionally subject to fire, or partially fall to fire, but the number and severity is greatly reduced due to proactive management, beefed up safety and evacuation protocols, and the restrictions on WUI expansion. Moderated climate warming makes wildfire management challenging, but not impossible. And beyond the WUI, the Green New Deal enables the funding and training of workers dedicated to land-fire stewardship and expansion of diverse fire design strategies across the state.

In the above scenario, the public plays a pivotal role in fostering needed change from the status quo, mainly by effectively demanding it. Likewise, large organizations and institutions (both governmental and non-governmental) can have a decisive role in how we collectively reimagine our relationship to fire, and address related past injustices and colonial legacies. Organizations have the collective power to do this by revisiting and refining their mission,

Figure 7.3 (*Right*)
A map of UC Reserves (in black) spread across all major ecosystems of California (in green) and many indigenous territories (outlined in white).

shifting how they operate, and by listening and responding to their constituencies. Given the vastly different ways organizations and institutions are structured and the kinds of work they do, they offer different kinds of possibilities. Public universities, like the one we (the authors) work for, and federal agencies like the U.S. Forest Service, are just two examples we might explore.

## Right to Burn

The University of California is a massive, multi-campus, R1/top-tier public research institution. It bills itself as an international leader in research innovation, teaching, diversity and sustainability. But its lofty claims came into considerable question beginning around 2020. As a land grant institution, UC was originally gifted 150,000 acres of land appropriated from Native Americans. Many now refer to these institutions as "land grab" universities that, in total, were granted about 11 million acres.[40] A two-day conference held at UC Berkeley around this time took this topic head on, exploring how that land grab is "intricately tied to California's unique history of Native dispossession and genocide" and examined "how UC continues to benefit from this wealth accumulation today."[41] Discussions were held and recommendations were made for what the university could do to try to address its responsibility to indigenous tribes.

At 756,000 acres, the UC reserves are one of the largest in the U.S. and encompass all major ecosystems in the state. They are spread across California's Floristic Province's biodiversity hotspots. Right to Burn is the initiative started in the mid-2020s that re-envisions the University of California Reserve System as a broad network of experimental and educational public landscapes for the reparative and creative reintroduction of fire and indigenous land stewardship and rights as a whole. It transforms the role of these lands from passive scientific observation to one of proactive adaptation, eco-cultural reparation, and public education.

As reparation, this initiative invites local native tribes near UC reserve lands to re-assert their rights to cultural burns and other traditional ecological practices within them. They are also invited to assume jobs (and to rewrite job descriptions) as full-time reserve staff and to co-lead and advise on research and educational initiatives. These hybrid forms of landscape stewardship (Western science and traditional ecological knowledge) are brought together with diverse UC faculty, student research programs and work internships to generate pluralistic and useful forms of knowledge that spread beyond the UC reserves to adjacent public lands, based on their results and findings. As a shared, experimental place for learning (socially, politically, and ecologically), U.S. and international students gain first-hand experience of indigenous ways of tending the land. Tribes and cultural burners are able to observe how their practices inform Western science and expand the domains of its research and vice versa.

The reserves become state-of-the-art landscape labs for land-fire stewardship. And rather than a closed-door policy, public outreach and engagement across the reserves informs larger publics and catalyzes change far beyond their borders. People come from around the world to study and learn from this initiative, from chancellors to community activists.

Right to Burn memes its way into the U.S. Forest Service, who build relationships with tribes and cultural burning practitioners to burn and steward lands on ancestral

Figure 7.4 (*Right*) Two potential futures for the Putah Creek Riparian Reserve – one that is unmanaged (left) and one that is carefully tended using traditional ecological knowledge and cultural burning to rehabilitate the landscape (right).

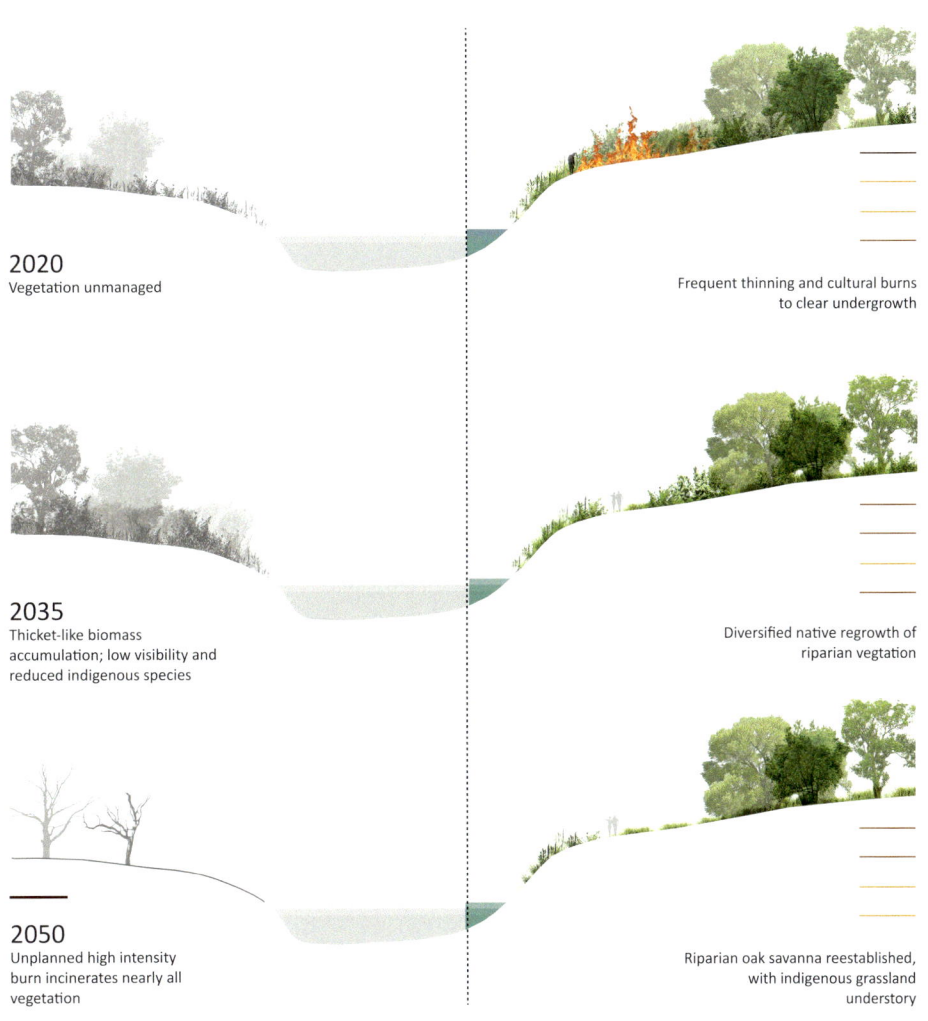

**2020**
Vegetation unmanaged

Frequent thinning and cultural burns
to clear undergrowth

**2035**
Thicket-like biomass
accumulation; low visibility and
reduced indigenous species

Diversified native regrowth of
riparian vegtation

**2050**
Unplanned high intensity
burn incinerates nearly all
vegetation

Riparian oak savanna reestablished,
with indigenous grassland
understory

50 ft

*homelands within their agency's domain. This starts through small-scale pilot projects that demonstrate their value to both parties, and then spreads and expands.[42] This relationship grows to be a symbiotic and power-shared relationship. Tribal members are provided with well-paying jobs for their services and benefits and offer cultural burn training to the agencies' staff. For the first time in decades, the Forest Service becomes successful at more sustainably managing the lands charged to them, and in ways that create a wider range of benefits to people and other organisms. In turn, shared land right agreements are made and the forest service becomes a radically different public agency. This, in turn, leads to further change in federal and state agencies.*

*As of 2047 the Right to Burn initiative employs over 60 indigenous Americans, fosters the establishment of entirely new majors and academic programs, and creates a range of national and international research collaborations with other universities, public agencies, and other institutions. Like all changes and evolutions in cultural, social, and technical norms, mistakes, misunderstandings and differences in participant expectations are common in the early stages of these efforts and are part of learning and experimenting with new ways of living and coexisting. The initiative's foresight and planning for those difficulties and its ability to proactively and openly address them, is a major component in the success of the experiment.*

## Closing

We don't know what future scenarios will be the ones we actually come to experience and embody, and almost certainly, none of them will be exactly like the ones above. We can't know the degree to which indigenous sovereignty and cultural burning will grow. We don't know how land-fire stewardship will be fostered through funding, policies, and education. We don't know how much and how quickly our climate and ecosystems are going to change, nor what the nature of fire will be because of those changes. But envisioning exploratory scenarios of what could possibly happen (like *Wrath of Fire*), combined with more normative scenarios of what we might want to make happen (based on current values and concerns), may provide us with broader options and horizons to consider. The scenarios illustrated here are just a beginning gesture to prompt us to think and feel more expansively and inclusively about the kinds of relationships we want to have with fire, and from which we can set about the work of designing and enacting those relations.

Whatever pyro futures come to be, they will likely consist of blends of resistance, co-creation, and retreat approaches, varied across situations and contexts. Given thousands of years of collective human experience with fire, we know we are not separable from it. It has remade us and we have remade it, based on how we act. In gleaning from the past, we can see just how different human relationships with fire can be, and how consequential and formative those differences are for landscapes. Landscapes change responsively to fire. Any relationships we now make with fire – even if it attempts to be completely hands off – will still bear a heavy imprint of our agency, which is why design by, and with fire is so important. We have, and always have had, so much choice in what nascent, fiery landscapes can be.

# Notes

1   Francisco Moreira, Davide Ascoli, Hugh Safford, Mark A. Adams, José M. Moreno, José M.C. Pereira, Filipe X. Catry et al., "Wildfire Management in Mediterranean-Type Regions: Paradigm Change Needed," *Environmental Research Letters* 15, no.1 (2020): 011001.

2   Anderson, M. Kat, *Tending the Wild: Native American Knowledge and the Management of California's Natural Resources*. Oakland: University of California Press, 2013.

3   A. Petryna, "Wildfires at the Edges of Science: Horizoning Work Amid Runaway Change," *Cultural Anthropology* 33, no.4 (2018): 570–595.

4   Ibid (p. 585).

5   Ibid (p. 571).

6   Petryna, A. (2018).

7   Ibid (p. 579).

8   For example, see Angela Wilkinson, and Roland Kupers, "Living in the Futures," *Harvard Business Review* 91, no.5 (2013): 118–127.
    Lauren Rickards, Ray Ison, Hartmut Fünfgeld, and John Wiseman, "Opening and Closing the Future: Climate Change, Adaptation, and Scenario Planning," *Environment and Planning C: Government and Policy* 32, no.4 (2014): 587–602.
    Thomas J. Chermack and Laura M. Coons, "Integrating Scenario Planning and Design Thinking: Learnings from the 2014 Oxford Futures Forum," *Futures* 74 (2015): 71–77.

9   Paul F. Hessburg, Susan J. Prichard, R. Keala Hagmann, Nicholas A. Povak, and Frank K. Lake, "Wildfire and Climate Change Adaptation of Western North American Forests: A Case for Intentional Management," *Ecological Applications* 31, no.8 (2021): e02432.
    Jonathan Thompson, "The West's Forever Fire Season: How Climate Change Makes Wildfire More Likely to Happen All Year Round." *High Country News*, https://www.hcn.org/issues/54.8/infographic-the-wests-wildfire-forever-fire-season
    A. Park Williams, Benjamin I. Cook, and Jason E. Smerdon, "Rapid Intensification of the Emerging Southwestern North American Megadrought in 2020–2021," *Nature Climate Change* 12, no.3 (2022): 232–234.

10  Don Hankins, "Beyond the Plume of Smoke," Bay Nature, https://baynature.org/article/beyond-the-plume-of-smoke/

11  "Climate Change Indicators: Snowpack," EPA, https://www.epa.gov/climate-indicators/climate-change-indicators-snowpack

12  Rhoades, Alan M., Andrew D. Jones, and Paul A. Ullrich, "The Changing Character of the California Sierra Nevada as a Natural Reservoir," *Geophysical Research Letters* 45, no.23 (2018): 13008–13019.

13  Stephens, Scott L., Neil Burrows, Alexander Buyantuyev, Robert W. Gray, Robert E. Keane, Rick Kubian, Shirong Liu et al., "Temperate and Boreal Forest Mega-Fires: Characteristics and Challenges," *Frontiers in Ecology and the Environment* 12, no.2 (2014): 115–122.

14  Stephenson, Nathan, Christy Brigham, Sue Cag, Conservationist Anthony Caprio, Joshua Flickinger, Linnea Hardlund, Rodney Hart et al., "Preliminary Estimates of Sequoia Mortality in the 2020 Castle Fire" (2021).

15  Coop, Jonathan D., Sean A. Parks, Camille S. Stevens-Rumann, Shelley D. Crausbay, Philip E. Higuera, Matthew D. Hurteau, Alan Tepley et al., "Wildfire-Driven Forest Conversion in Western North American Landscapes," *BioScience* 70, no.8 (2020): 659–673.

16  Mowery, M., Anna Read, Kelly Johnston, and Tareq Wafaie, "Planning the Wildland-Urban Interface," PAS Report 594 (2019): 194.

17  Ibid.

18  Stephens, Scott L., Robert E. Martin, and Nicholas E. Clinton, "Prehistoric Fire Area and Emissions from California's Forests, Woodlands, Shrublands, and Grasslands," *Forest Ecology and Management* 251, no.3 (2007): 205–216.

19  Yamond Zhong, "How to Save a Forest by Burning It," *The New York Times*, https://www.nytimes.com/2022/09/07/climate/california-wildfire-prescribed-burn.html
    Hessburg, "Wildfire and Climate."

20  Stephens, Scott L., Jason J. Moghaddas, Carl Edminster, Carl E. Fiedler, Sally Haase, Michael Harrington, Jon E. Keeley et al., "Fire Treatment Effects on Vegetation Structure, Fuels, and Potential Fire Severity in Western US Forests," *Ecological Applications* 19, no.2 (2009): 305–320.
    McIver, James, Andrew Youngblood, and Scott L. Stephens, "The National Fire and Fire Surrogate Study: Ecological Consequences of Fuel Reduction Methods in Seasonally Dry Forests," *Ecological Applications* 19, no.2 (2009): 283–284.

21  *California's Strategic Plan for Expanding Beneficial Fire*, California's Wildlfire and Forest Resilience Task Force (2022), https://www.fire.ca.gov/media/xcqjpjmc/californias-strategic-plan-for-expanding-the-use-of-beneficial-fire-march-16_2022.pdf

22  Ibid.

23  Forest Stewards Guild, Insights and Suggestions for Certified Prescribed Burn Manager Programs, (2020), https://foreststewardsguild.org/wp-content/uploads/2020/03/InsightsRecommendationsCPMBprograms.pdf
    Weir, John R., Urs P. Kreuter, Carissa L. Wonkka, Dirac Twidwell, Dianne A. Stroman, Morgan Russell, and Charles A. Taylor, "Liability and Prescribed Fire: Perception and Reality," *Rangeland Ecology & Management* 72, no.3 (2019): 533–538.

24   M. Burke et al., "Managing the Growing Cost of Wildfire" Stanford Institute for Economic Policy Research, Policy Brief (October 2020), https://siepr.stanford.edu/publications/policy-brief/managing-growing-cost-wildfire

25   Kanawha Lake, Frank, "Fire as Medicine: Fire Dependent Cultures and Re-Empowering American Indian Tribes" (2018), https://fireadaptednetwork.org/fire-as-medicine-fire-dependent-cultures/

26   Anderson, Kat, *Tending the Wild: Native American Knowledge and the Management of California's Natural Resources* (Berkeley: University of California Press, 2005).

27   Adlam, Christopher, Diana Almendariz, Ron W. Goode, Deniss J. Martinez, and Beth Rose Middleton. "Keepers of the Flame: Supporting the Revitalization of Indigenous Cultural Burning." *Society & Natural Resources* 35, no.5 (2022): 575–590.

28   Ibid.

29   Clark, S. A., A. Miller, and D. L. Hankins. 2021. Good Fire: Current Barriers to the Expansion of Cultural Burning and Prescribed Fire in California and Recommended Solutions. Karuk Tribe, https://karuktribeclimatechangeprojects.com/good-fire/

30   Hilary Beaumont, "New California Law Affirms Indigenous Right to Controlled Burns," *Aljazeera*, https://www.aljazeera.com/news/2021/12/3/new-california-law-affirms-indigenous-right-to-controlled-burns

31   Red Nation. *The Red Deal: Indigenous Action to Save Our Earth*. Common Notions, 2021.

32   Daniel Godwin, Prescribed Fire Liability Insurance: Unavailable, Unaffordable, And Vital, The Ember Alliance, https://emberalliance.org/2022/06/28/prescribed-fire-liability-insurance-unavailable-unaffordable-and-vital/

33   SB-332 Civil Liability: Prescribed Burning Operations: Gross Negligence, California Legislative Information, https://leginfo.legislature.ca.gov/faces/billTextClient.xhtml?bill_id=202120220SB332

34   Ibid.

35   It limits liability to *gross negligence*, which can be generally defined as extreme and obvious negligence, demonstrable through ample evidence. See Varner, J. Morgan, J. Kevin Hiers, Slaton B. Wheeler, John McGuire, Lenya Quinn-Davidson, William E. Palmer, and Laurie Fowler. "Increasing Pace and Scale of Prescribed Fire via Catastrophe Funds for Liability Relief." *Fire* 4, no.4 (2021): 77.

36   AB-642 Wildfires. California Legislative Information, https://leginfo.legislature.ca.gov/faces/billNavClient.xhtml?bill_id=202120220AB642

37   For a deeper discussion on the role of experience and aesthetics in fostering intentional public change, see Rob Holmes, Brett Milligan, Gena Wirth, "Aesthetics and Publics" in Silt Sand Slurry: Dredging, Sediment, and the Worlds We Are Making, ORO Applied Research and Design Publishing, 2023.

38   For plausible examples of how collective revolt like this could happen, see Kim Stanley Robinson's *Ministry of the Future* (Fiction) and Naomi Klein's *On Fire*.

39   "Fire Adapted Communities," Fire Adapted Communities, https://fireadapted.org/

40   Lee, Robert, Tristan Ahtone, and Margaret Pearce. "Land-Grab Universities." *High Country News* 30 (2020).

41   "UC Land Grab," UC Berkeley Center for Educational Justice and Community Engagement, https://cejce.berkeley.edu/centers/native-american-student-development/uc-land-grab

42   Collaborations and pilot projects like this are well under way. For example, see: Long, Jonathan W., Frank K. Lake, Ron W. Goode, and Benrita Mae Burnette, "How Traditional Tribal Perspectives Influence Ecosystem Restoration," *Ecopsychology* 12, no.2 (2020): 71–82. Long, Jonathan W., and Frank K. Lake, "Escaping Social-Ecological Traps through Tribal Stewardship on National Forest Lands in the Pacific Northwest, United States of America," *Ecology and Society* 23, no.2 (2018). Sowerwine, Jennifer, Daniel Sarna-Wojcicki, Megan Mucioki, Lisa Hillman, Frank Lake, and Edith Friedman, "Enhancing Food Sovereignty: A Five-Year Collaborative Tribal-University Research and Extension Project in California and Oregon," *Journal of Agriculture, Food Systems, and Community Development* 9, no.B (2019): 167–190.

# Bibliography

AB-642 Wildfires. California Legislative Information. https://leginfo.legislature.ca.gov/faces/billNavClient.xhtml?bill_id=202120220AB642

Adlam, Christopher, Diana Almendariz, Ron W. Goode, Deniss J. Martinez, and Beth Rose Middleton. "Keepers of the Flame: Supporting the Revitalization of Indigenous Cultural Burning." *Society & Natural Resources* 35, no.5 (2022): 575–590.

Anderson, Kat. *Tending the Wild: Native American Knowledge and the Management of California's Natural Resources*. Berkeley: University of California Press, 2005.

Beaumont, Hilary. "New California Law Affirms Indigenous Right to Controlled Burns." *Aljazeera*. https://www.aljazeera.com/news/2021/12/3/new-california-law-affirms-indigenous-right-to-controlled-burns

Burke et al., "Managing the Growing Cost of Wildfire" Stanford Institute for Economic Policy Research, Policy Brief (October 2020). https://siepr.stanford.edu/publications/policy-brief/managing-growing-cost-wildfire

*California's Strategic Plan for Expanding Beneficial Fire*, California's Wildfire and Forest Resilience Task Force. (2022). https://www.fire.ca.gov/media/xcqjpjmc/californias-strategic-plan-for-expanding-the-use-of-beneficial-fire-march-16_2022.pdf

Chermack, Thomas J., and Laura M. Coons. "Integrating Scenario Planning and Design Thinking: Learnings from the 2014 Oxford Futures Forum." *Futures* 74 (2015): 71–77.

Clark, Sara A., Andrew Miller, and Don L. Hankins. Good Fire: Current Barriers to the Expansion of Cultural Burning and Prescribed Fire in California and Recommended Solutions. Karuk Tribe. (2021). https://karuktribe-climatechangeprojects.com/good-fire/

"Climate Change Indicators: Snowpack", EPA. https://www.epa.gov/climate-indicators/climate-change-indicators-snowpack

Coop, Jonathan D., Sean A. Parks, Camille S. Stevens-Rumann, Shelley D. Crausbay, Philip E. Higuera, Matthew D. Hurteau, Alan Tepley et al. "Wildfire-Driven Forest Conversion in Western North American Landscapes." *BioScience* 70, no.8 (2020): 659–673.

"Fire Adapted Communities", Fire Adapted Communities. https://fireadapted.org/

Forest Stewards Guild. Insights and Suggestions for Certified Prescribed Burn Manager Programs. (2020). https://foreststewardsguild.org/wp-content/uploads/2020/03/InsightsRecommendationsCPMBprograms.pdf

Godwin, Daniel, Prescribed Fire Liability Insurance: Unavailable, Unaffordable, and Vital, The Ember Alliance. https://emberalliance.org/2022/06/28/prescribed-fire-liability-insurance-unavailable-unaffordable-and-vital/

Hankins, Don, "Beyond the Plume of Smoke" Bay Nature. https://baynature.org/article/beyond-the-plume-of-smoke/

Hessburg, Paul, Susan J. Prichard, R. Keala Hagmann, Nicholas A. Povak, and Frank K. Lake. "Wildfire and Climate Change Adaptation of Western North American Forests: A Case for Intentional Management." *Ecological applications* 31, no.8 (2021): e02432.

Kanawha Lake, Frank, "Fire as Medicine: Fire Dependent Cultures and Re-Empowering American Indian Tribes." (2018). https://fireadaptednetwork.org/fire-as-medicine-fire-dependent-cultures/

Lee, Robert, Tristan Ahtone, and Margaret Pearce. "Land-Grab Universities." *High Country News* 30 (2020).

Long, Jonathan W., Frank K. Lake, Ron W. Goode, and Benrita Mae Burnette. "How Traditional Tribal Perspectives Influence Ecosystem Restoration." *Ecopsychology* 12, no.2 (2020): 71–82.

Long, Jonathan W., and Frank K. Lake. "Escaping Social-Ecological Traps through Tribal Stewardship on National Forest Lands in the Pacific Northwest, United States of America." *Ecology and Society* 23, no.2 (2018).

McIver, James, Andrew Youngblood, and Scott L. Stephens. "The National Fire and Fire Surrogate Study: Ecological Consequences of Fuel Reduction Methods in Seasonally Dry Forests." *Ecological Applications* 19, no.2 (2009): 283–284.

Moreira, Francisco, Davide Ascoli, Hugh Safford, Mark A. Adams, José M. Moreno, José M.C. Pereira, Filipe X. Catry et al. "Wildfire Management in Mediterranean-Type Regions: Paradigm Change Needed." *Environmental Research Letters* 15, no.1 (2020): 011001.

Mowery, M., Anna Read, Kelly Johnston, and Tareq Wafaie. "Planning the Wildland-Urban Interface" PAS Report 594. (2019): 194.

Petryna, A., "Wildfires at the Edges of Science: Horizoning Work Amid Runaway Change." *Cultural Anthropology* 33, no.4 (2018): 570–595.

Red Nation. *The Red Deal: Indigenous Action to Save Our Earth*. Common Notions, 2021.

Rhoades, Alan M., Andrew D. Jones, and Paul A. Ullrich. "The Changing Character of the California Sierra Nevada as a Natural Reservoir." *Geophysical Research Letters* 45, no.23 (2018): 13008–13019.

Rickards, Lauren, Ray Ison, Hartmut Fünfgeld, and John Wiseman. "Opening and Closing the Future: Climate Change, Adaptation, and Scenario Planning." *Environment and Planning C: Government and Policy* 32, no.4 (2014): 587–602.

SB-332 Civil Liability: Prescribed Burning Operations: Gross Negligence, California Legislative Information. https://leginfo.legislature.ca.gov/faces/billTextClient.xhtml?bill_id=202120220SB332

Sowerwine, Jennifer, Daniel Sarna-Wojcicki, Megan Mucioki, Lisa Hillman, Frank Lake, and Edith Friedman. "Enhancing Food Sovereignty: A Five-Year Collaborative Tribal-University Research and Extension Project in California and Oregon." *Journal of Agriculture, Food Systems, and Community Development* 9, no.B (2019): 167–190.

Stephens, Scott L., Neil Burrows, Alexander Buyantuyev, Robert W. Gray, Robert E. Keane, Rick Kubian, Shirong Liu et al. "Temperate and Boreal Forest Mega-Fires: Characteristics and Challenges." *Frontiers in Ecology and the Environment* 12, no.2 (2014): 115–122.

Stephens, Scott L., Robert E. Martin, and Nicholas E. Clinton. "Prehistoric Fire Area and Emissions from California's Forests, Woodlands, Shrublands, and Grasslands." *Forest Ecology and Management* 251, no.3 (2007): 205–216.

Stephens, Scott L., Jason J. Moghaddas, Carl Edminster, Carl E. Fiedler, Sally Haase, Michael Harrington, Jon E. Keeley et al. "Fire Treatment Effects on Vegetation Structure, Fuels, and Potential Fire Severity in Western US Forests." *Ecological Applications* 19, no.2 (2009): 305–320.

Stephenson, Nathan, Christy Brigham, Sue Cag, Conservationist Anthony Caprio, Joshua Flickinger, Linnea Hardlund, Rodney Hart et al. "Preliminary Estimates of Sequoia Mortality in the 2020 Castle Fire." (2021).

Thompson, Jonathan, "The West's Forever Fire Season: How Climate Change Makes Wildfire More Likely to Happen All Year Round." *High Country News*, https://www.hcn.org/issues/54.8/infographic-the-wests-wildfire-forever-fire-season

"UC Land Grab", UC Berkeley Center for educational Justice and Community Engagement. https://cejce.berkeley.edu/centers/native-american-student-development/uc-land-grab

Varner, J. Morgan, J. Kevin Hiers, Slaton B. Wheeler, John McGuire, Lenya Quinn-Davidson, William E. Palmer, and Laurie Fowler. "Increasing Pace and Scale of Prescribed Fire via Catastrophe Funds for Liability Relief." *Fire* 4, no.4 (2021): 77.

Weir, John R., Urs P. Kreuter, Carissa L. Wonkka, Dirac Twidwell, Dianne A. Stroman, Morgan Russell, and Charles A. Taylor. "Liability and Prescribed Fire: Perception and Reality." *Rangeland Ecology & Management* 72, no.3 (2019): 533–538.

Wilkinson, Angela and Roland Kupers. "Living in the Futures." *Harvard Business Review* 91, no.5 (2013): 118–127.

Williams, A. Park, Benjamin I. Cook, and Jason E. Smerdon. "Rapid Intensification of the Emerging Southwestern North American Megadrought in 2020–2021." *Nature Climate Change* 12, no.3 (2022): 232–234.

Zhong, Yamond, "How to Save a Forest by Burning It". *The New York Times*, https://www.nytimes.com/2022/09/07/climate/california-wildfire-prescribed-burn.html

Figure 8.1
View of surviving trees in valleys around Lake Berryessa following the LNU Lightning Complex Fires of 2020.

Figure 8.2
Close-up view of a tree grove after the LNU Lightning Complex Fires of 2020.

Figure 8.3
View of a dense conifer forest. Photograph by Derek Young.

Figure 8.4
View of Quail Ridge Reserve near Lake Berryessa following the LNU Lightning Complex Fires of 2020.

Figure 8.5
The effects of a firenado in the 2018 Carr Fire. Photograph by Malcolm North.

Figure 8.6
View of a valley near Lake Berryessa following the LNU Lightning Complex Fires of 2020.

Figure 8.7
View of a conifer grove after a wildfire event with patchy tree survival. Photograph by Derek Young.

# Acknowledgments

This book would not have been possible without the help and support of many individuals. First, and foremost, we would like to thank our student research assistants – Virginia Morgan, Aiyuan Liao, and Jay Deisman – for helping to frame out the project in its early stages. We'd also like to extend our gratitude to Andrew Latimer and Stephen Wheeler for their expertise in fire ecology and urban planning, and their willingness to serve as sounding boards and active brainstormers for the project. To the editors Stacy Morford at *The Conversation* and Jiayi Zhou at *Landscape Architecture Frontiers*, thank you for your detailed feedback and suggestions regarding the design case studies. Gavin Kroeber, Arleene Correa, and the rest of *Fire School* – thank you for the book discussions, burn zone field trips, and the always-engaging conversations. We would also like to thank all of our interviewees – Margo Robbins, Val Charlton, Dean Turner, Gerard Jadoul, Kerry Metlen, Alanna Rebelo, Marta Carola, Garrett Dickman, Dan Buckley, and Scott Gediman – for carving time out of their busy schedules to talk with us about their fire stewardship work. Thank you to Don Hankins for many conversations about fire and ecocultural restoration. Thank you also to Hanna Prissen, Jordan Duke, and Sarah Toth for their bold ideas on how we might better live with fire. Hugh Safford, Malcolm North, Michael Koontz, and Derek Young – thank you for sharing your fire ecology fieldwork imagery and insights. For the folks at Routledge – Grace Harrison, Kate Schell, Russell George, and Megha Patel – thank you for believing in us, providing guidance, and steering us toward the finish line. Lastly, we would like to thank our families and friends for supporting us on this book-writing adventure.

# Figures

# Index

Note: *Italic* page numbers refer to figures.